# The Measurement of Visual Motion

**ACM Distinguished Dissertations**

1982

**Abstraction Mechanisms and Language Design**
by Paul N. Hilfinger

**Formal Specification of Interactive Graphics Programming Languages**
by William R. Mallgren

**Algorithmic Program Debugging**
by Ehud Y. Shapiro

1983

**The Measurement of Visual Motion**
by Ellen Catherine Hildreth

**Synthesis of Digital Designs from Recursion Equations**
by Steven D. Johnson

# The Measurement of Visual Motion

Ellen Catherine Hildreth

The MIT Press
Cambridge, Massachusetts
London, England

This book was printed and bound in the United States of America.

Publisher's Note: This format is intended to reduce the cost of publishing certain works in book form and to shorten the gap between editorial preparation and final publication. Detailed editing and composition have been avoided by photographing the text of this book directly from the author's word-processor output.

Dissertation submitted to the Department of Electrical Engineering and Computer Science, Massachusetts Institute of Technology, May 1983.

**Library of Congress Cataloging in Publication Data**

Hildreth, Ellen Catherine.
The measurement of visual motion.

(ACM distinguished dissertations)
Thesis (Ph. D.) – Massachusetts Institute of Technology, 1983.
Bibliography: p.
Includes index.
1. Motion perception (Vision) – Data processing.
2. Digital computer simulation. I. Title. II. Series.
BF241.H53 1984 153.7'54 83-25112
ISBN 0-262-08143-1

## Series Foreword

This book is being published by The MIT Press as an outgrowth of the annual contest for the best doctoral dissertation in computer-related science and engineering. The contest was initiated in 1982 and is co-sponsored by ACM and The MIT Press.

The Distinguished Doctoral Dissertation Series has been created to recognize that some of the theses considered in the final round of selecting a contest winner also deserve publication. In the judgment of the ACM selection committee and The MIT Press, this thesis is of such high quality that it deserves special recognition in this new series.

Dr. Ellen Hildreth wrote her thesis on "The Measurement of Visual Motion" at MIT. The thesis work was supervised by Dr. Shimon Ullman, MIT Professor of Psychology, and the thesis was submitted to the 1983 competition. The Doctoral Dissertation Award Committee of ACM recommended its publication because it blends a comprehensive grasp of disparate scientific and engineering disciplines. Both theoretical analysis and experimental validation are well presented. This work substantially advances computer-based techniques for measuring and understanding visual phenomena in the human eye.

After completing her dissertation, Dr. Hildreth prepared a revised version of parts of the thesis. These revisions have been included in this book.

Walter M. Carlson
Chairman, ACM Awards Committee

## Acknowledgments

Many people have contributed to this work and provided advice and encouragement to its author. I would like to express my gratitude to:

My supervisor, Shimon Ullman, for his generous support and intellectual guidance throughout my years as a graduate student; his insightful advice, clear vision, and patience to see the work through its more tenuous stages, were invaluable;

Tomaso Poggio, for his personal support and encouragement, his sound technical advice, and for constantly pushing me forward by always asking the next question;

Whitman Richards, whose encouragement and enthusiasm kept me thinking that there really was a light at the end of the tunnel, and who often helped to broaden my work with a new and interesting perspective;

Berthold Horn, who provided valuable technical comments, and suggested one of the key ideas on which this work is based; and who, in ways both large and small, helped me through my graduate career;

Tomas Lozano-Perez, for many valuable comments on drafts of this book, and for helping to create a supportive and friendly environment, by giving so much of himself to those around him;

Patrick Winston, for his financial and intellectual support, and for the stimulating research environment that he maintains at the MIT Artificial Intelligence Laboratory;

Members of the Vision Group, for serving as a valuable source for comments and criticisms; I especially thank Eric Grimson and Alan Yuille, who provided key insights that helped to place this work on a firmer theoretical foundation;

David Marr, who first encouraged me to enter the field of vision, who provided constant and generous support and guidance as I took my first uneasy steps, and whose memory will continue to be a source of inspiration throughout my career;

Lorraine Gray, Matt and Mary Mason, and Dan Brotsky, whose friendship helped to make this experience much more enjoyable;

Eric Grimson, who shared with me the pains and happiness of the last few years; his endless support kept me going through it all;

Tammie, Ruffian, Amy and Morgan Le Fay, who kept reminding me that there was more to life than writing a PhD thesis;

and finally, to my parents, who first taught me the importance of education.

## Table of Contents

# The Measurement of Visual Motion

# 1. INTRODUCTION

## 1.1. The Problem and Approach

Although the world with which we interact contains complex three-dimensional surfaces that are moving in space, the projection of light onto the eye presents our visual system with only a two-dimensional image of the world, that is changing with time. One of the functions of vision is to reconstruct a three-dimensional representation of the world from this two-dimensional image. A primary source of information, from which this reconstruction may take place, is given by the organization of movement in the changing image. The motion of elements in the image is not provided to our visual system directly, however; it must be inferred from the pattern of changing intensity that reaches the eye. This book presents a study of early motion analysis, whose focus is the measurement of motion in the changing image, and the use of relative movement for locating surface boundaries. A primary motivation of this work is to understand the way in which this analysis takes place in the human visual system.

The approach that is taken in this study is a *computational* one (Marr 1976a, 1976b, 1982; Marr and Poggio 1977). While our main interest is to understand an aspect of human vision, a basic tenet of the computational paradigm is that the human system may be viewed as an *information processor*, which performs computations on internal symbolic representations of visual information. There is a level of description of the computations performed by the human system that is shared with any visual information processor, which David Marr referred to as the *computational theory*. This level of description first defines the goal of a visual processing task. Perceptual studies contribute to the formulation of this goal, by establishing the feasibility of carrying out the particular task. For example, the feasibility of recovering the three-dimensional structure of surfaces from their motion has been demonstrated in many perceptual studies (for example, Wallach and O'Connell 1953; Johansson 1973, 1975; Ullman 1979a). A second aim of the computational theory is to make explicit the underlying physical assumptions that allow a particular computation to succeed. The measurement of motion, like other early visual processing problems, is insoluble from information provided by the changing image alone. Additional constraint, in the form of basic assumptions about the physical world that generally hold true, is necessary to restrict the problem sufficiently to allow its solution. In general, the choice of underlying assumptions for a particular visual task may not be unique. The aim of the present work is to discover the particular assumptions that underly the measurement of motion in the human visual system; fortunately, different choices for these assumptions often fail in characteristic ways that can be examined through perceptual studies.

After formulating the computational theory for a visual processing problem, we then consider particular *representations* for the input and output of the computation, and an *algorithm* for transforming one representation into the other. This level of analysis is tied more closely to the particular visual system that we seek to understand. Both perceptual and neurophysiological studies provide insight into the particular representations and algorithm that underly the measurement of motion in the human visual system. In general, we are interested in algorithms that are biologically feasible, suggesting that they involve simple, local, parallel operations (Ullman 1979b; Grimson 1981; Marr 1982). Finally, the implementation of an algorithm for measuring visual motion provides a tool for testing the adequacy of the computational theory. It allows a rigorous examination of the validity of the constraints imposed on the computation, and the physical plausibility of its solution. Through the performance of the algorithm on known perceptual demonstrations, the predictions of the theory may be compared with human motion perception.

## 1.2. The Measurement of Visual Motion

The analysis of visual motion divides naturally into two stages. The first is the measurement of motion in the changing retinal image; the second is the use of motion measurements, for example, to separate the scene into distinct objects, and infer their three-dimensional structure. In this study, we assume that these two stages may be performed independently, and focus on the first stage: the measurement of visual motion. The goal of this measurement is formulated as the computation of a two-dimensional velocity field that assigns a direction and speed of movement to elements in the changing image (similar to the *optical flow field* described by Gibson (1950)). Figure 1.1 illustrates a velocity field that was computed from a pair of aerial photographs, taken in sequence from an airplane. In the background, the first image is shown with reduced contrast. The superimposed black lines represent local velocity vectors that describe the movement of features in the image, as the

**Figure 1.1. A Velocity Field.** The black lines represent local velocity vectors, displayed at evenly spaced points over the original image.

plane flies over the scene. For some visual tasks, it may be sufficient to compute only certain properties of the velocity field; for example, to respond quickly to a moving object, motion must be detected, but not necessarily measured. Other tasks, such as the recovery of three-dimensional structure, require a more complete and accurate computation of the velocity field (Ullman 1979a, 1980, 1983a,b; Prazdny 1980, 1983; Clocksin 1980; Longuet-Higgins and Prazdny 1981; Longuet-Higgins 1981).

The measurement of visual motion poses significant theoretical problems for a computational study. One problem, referred to by some researchers as the *aperture problem*, is that local measurements of movement in the changing image generally provide only one component of local velocity, rather than the full two-dimensional velocity vector (Wallach 1976; Fennema and Thompson 1979; Marr and Ullman 1981; Horn and Schunck 1981; Adelson and Movshon 1982). A second, deeper theoretical problem is that the motion of elements in the image is not determined uniquely by the pattern of changing intensity. In the remainder of this section, we describe these two problems in more detail, and briefly summarize the work on motion measurement presented in this book.

The aperture problem is illustrated in Figure 1.2a. Suppose the motion of the edge E is analyzed by a local movement detector that examines a limited area of the image, represented by the aperture A. Such a detector can measure only the component of motion in the direction perpendicular to the orientation of the edge, indicated by the vector **c**. The component of motion along the edge is invisible through this limited aperture. Thus, a local detector cannot distinguish between movement in the directions indicated by **b**, **c** and **d** in Figure 1.2a. In Figure 1.2b, a circle undergoes pure translation to the right, as indicated by the arrow at the center of the circle. The vectors along the contour represent the perpendicular components of velocity that can be measured directly from the changing image. These local measurements each provide some constraint on the possible motions of the circle. Its true motion, however, can be determined only by combining the constraints imposed by these local measurements. As a consequence of the aperture problem, the measurement of motion requires two stages of analysis: first, local motion measurements are obtained from the changing image, and generally provide only one component of velocity; second, these local measurements are combined to compute the full two-dimensional velocity field.

The measurement of motion may be performed at different stages in the processing of an image. For example, it could follow the initial registration of intensity, the detection of primitive features such as edges or intensity changes, or the delineation of the image into distinct objects (Ullman 1979a). Marr and Ullman (1981) proposed that in the human visual system, initial motion measurements may be made at the locations of significant intensity changes. This allows motion measurement to be performed early in the analysis of an image, in areas of the image where this measurement is most reliable. These intensity changes are detected in such a way that they form contours in the image, and the initial motion measurements provide the component of velocity in the direction perpendicular to the local orientation of the contours. In this study, we consider the work of

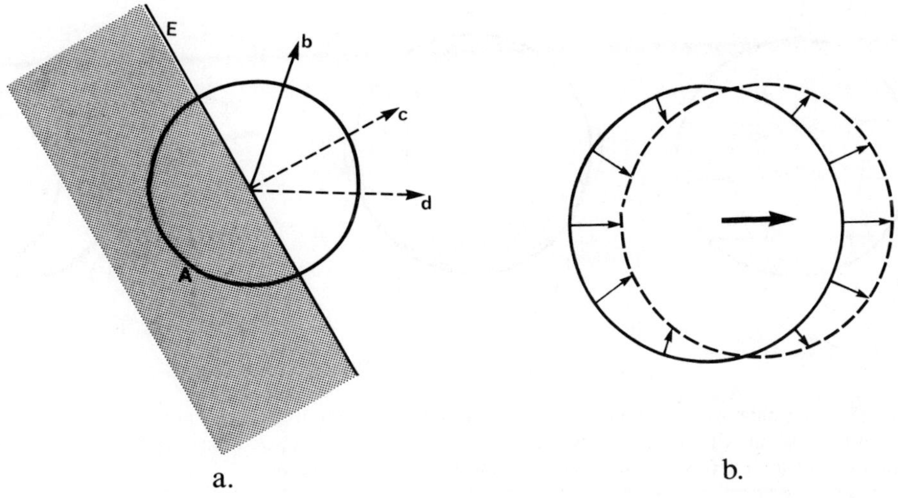


Figure 1.2. The Aperture Problem. (a) An operation that views the moving edge E through the local aperture A can compute only the component of motion c in the direction perpendicular to the edge. The true motion of the edge is ambiguous. (b) The circle undergoes pure translation to the right; the arrows represent the local perpendicular components of velocity obtained from the changing image.

Marr and Ullman as a starting point, and assume that the initial measurement of motion takes place along contours in the image, which correspond to the locations of significant intensity changes. We then examine the computation of the subset of the two-dimensional velocity field that exists along these contours.

A second theoretical problem that must be addressed in a computational study is that the motion of elements in the image is not determined uniquely from the pattern of changing intensity. Two factors contribute to this ambiguity of motion. The first is the loss of information due to the projection of the three-dimensional world onto a two-dimensional image; multiple surfaces, undergoing different motions in space, may give rise to the same two-dimensional image. The second factor is the loss of information due to the projection of physical surfaces in motion into a pattern of changing intensity. In general, the image that a surface projects onto the eye may not be sufficient to determine its movement in space.[1] One consequence of

[1] As an extreme example, a matte white sphere, rotating about a central axis, cannot be determined as such, on the basis of its projected image.

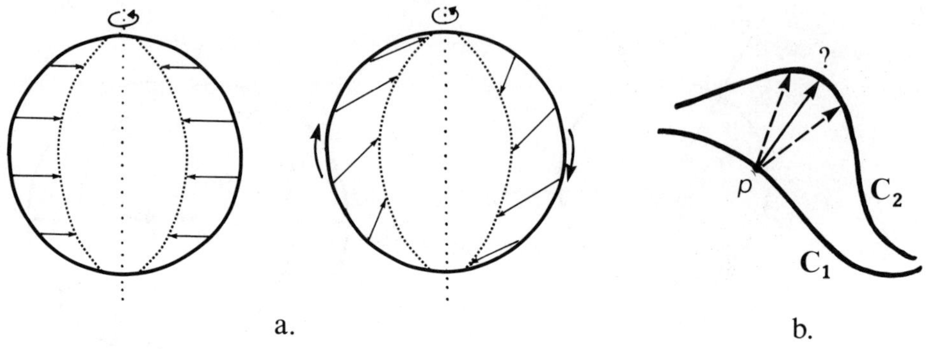


**Figure 1.3. Ambiguity of the Velocity Field.** (a) The arrows represent two possible velocity fields that are consistent with the changing image. (b) The curve $C_1$ rotates, translates and deforms over time to yield the curve $C_2$. The velocity of the point $p$ is ambiguous.

these two factors is that the two-dimensional velocity field that is computed from the changing image may not correspond to the true projected velocity field for the actual surfaces moving in space.

Figure 1.3 presents two simple examples that illustrate this ambiguity of motion in the image. In Figure 1.3a, the solid and dotted lines represent the image of a moving circle, at different instants of time. In the first frame (solid line), the circle lies parallel to the image plane, while in the second frame (dotted line), the circle is slanted in depth. One velocity field that is consistent with the two frames is derived from pure rotation of the circle about its central vertical axis, as shown to the left in Figure 1.3a. (The arrows represent a sample of the local velocity vectors along the circle.) There could also be a component of rotation in the plane of the circle, about its center, as shown to the right in Figure 1.3a. In this case, there are many possible velocity fields that one could assign to the changing image, which correspond to rigid motions of a circle in space. In addition, the changing image of Figure 1.3a might represent the projection of a different three-dimensional curve that is deforming over time, giving rise to yet another projected velocity field. In Figure 1.3b, the curve $C_1$ rotates, translates and deforms over time, to yield the curve $C_2$. The motion of points from $C_1$ to $C_2$ is again ambiguous (consider, for example, different possible velocities for the point $p$). In general, there are infinitely many two-dimensional velocity fields that are consistent with the changing image.

As a consequence of this ambiguity, the computation of a unique velocity field requires additional constraint, based on assumptions about the physical world that generally hold true. In this work, we rely upon the assumption that the physical world consists predominantly of solid objects, whose surfaces are generally smooth compared with their distance from the viewer. Under motion, a smooth surface usually generates a smoothly varying velocity field. Transforming this physical assumption into a computational constraint, we intuitively seek a velocity field that is consistent with the changing image, and varies smoothly. Unfortunately, there is an infinity of possible velocity fields that satisfy these two properties. Horn and Schunck (1981), in their work on the optical flow computation, suggest that a single solution may be obtained by finding the velocity field that varies the least, among the set of possible velocity fields consistent with the changing image.

In order to find the velocity field of least variation, some means of measuring the variation in velocity is required. Through a mathematical analysis, we find that by measuring the total change in the full two-dimensional velocity vector along contours in the image, this additional constraint can be formulated in a way that guarantees a unique solution to the velocity field computation. There are at least two classes of motion for which an algorithm utilizing this constraint necessarily computes the correct physical velocity field, assuming orthographic projection of the scene onto the image plane. The first consists of arbitrary rigid objects undergoing pure translation. The second consists of three-dimensional objects, whose edges are straight lines, undergoing rigid rotation and translation in space.

For the class of smooth curves, undergoing general motion in space, the velocity field of least variation is, in general, not the physically correct one. Through an empirical investigation, it is shown that the smoothest velocity field is often qualitatively similar to the true velocity field. Where the two deviate significantly, it appears that the smoothest velocity field may be more consistent with human motion perception. For example, the perception of radial motion when viewing a rotating spiral, or the perceived vertical motion of the stripes on a rotating barberpole, are more consistent with the velocity field of least variation, than the true velocity field.

In Chapter 2, some essential background material is presented, that led to this particular formulation of the motion measurement problem. Theoretical aspects of the velocity field computation are discussed in Chapter 3. Chapter 4 then describes

a. b.

**Figure 1.4. Dynamic Random Dot Pattern.** The two patterns in (a) and (b) are identical, except that a central patch of dots in (a) is shifted to the right in (b).

some algorithms that perform this computation, utilizing the additional constraint of smoothness of the velocity field. The performance of one particular algorithm is illustrated, for a number of motion sequences. This computational study raises many questions for the motion computation in the human visual system, which may be examined through perceptual studies. These questions are addressed in Chapter 6. Finally, implications of this study for the neurophysiological investigation of motion detection and measurement are addressed in Chapter 8.

### 1.3. The Computation of Surface Boundaries from Motion

If two adjacent surfaces undergo different motions, they generally give rise to a discontinuity in the velocity field along their boundary. The human visual system is capable of detecting these discontinuities, and utilizing them to identify object boundaries. A movie consisting of patterns of random dots, in which a central patch in the second frame is shifted relative to the first (see Figure 1.4), elicits a vivid impression of a central object separated from its background, with sharply defined boundaries (Anstis 1970; Regan and Spekreijse 1970; Julesz 1971; Braddick 1973, 1974). This demonstration suggests that the analysis of relative movement, and its use in establishing object boundaries, may proceed in the absence of all other cues to the organization of the scene. It raises the computational problem of explicitly detecting these discontinuities in the velocity field.

If it were possible to compute a precise local velocity field initially, the problem of detecting discontinuities might not be difficult. The goal would be to detect sudden changes in the direction or magnitude of velocity. We could, for example, apply techniques similar to those available for detecting changes in other image properties, such as intensity (Marr and Hildreth 1980). Without prior knowledge of surface boundaries, however, it is difficult to compute the velocity field correctly. This difficulty arises from the need to combine local motion measurements over an area of the image, which often leads to considerable error in the computation of the velocity field at object boundaries (Horn and Schunck 1981).

A second approach to the detection of discontinuities in the velocity field is to base this detection on more primitive motion measurements, such as those suggested by Marr and Ullman (1981). That is, the location of discontinuities is inferred prior to, or in parallel with, the combination of local motion measurements to compute the two-dimensional velocity field. The advantages of this approach are first, the detection can be fast, because it takes place at the earliest possible stage in the measurement of motion; and second, it aids the computation of the velocity field, by establishing the boundaries within which local measurements may be combined. It is also possible that a rough description of discontinuities is constructed from primitive motion measurements, and later refined after the full velocity field computation.

In Chapter 5, the detection of motion discontinuities is addressed from a computational viewpoint. One simple algorithm is presented, which is based on deriving discontinuities from the initial motion measurements along contours in the image. The performance of this algorithm on dynamic random dot patterns supports the feasibility of this approach to discontinuity detection. There has been considerable psychophysical work on the detection of motion discontinuities in the human visual system (for example, Lee 1972; Braddick 1973, 1974, 1980; Bell and Lappin 1973; Petersick, Hicks and Pantle 1978; Baker and Braddick 1982a,b; Nakayama 1981; Hadani, Gur and Meiri 1981; Van Doorn and Koenderink 1982, 1983; Chang and Julesz 1983). In Chapter 7, this work is incorporated into a discussion of specific questions that arise for the computations underlying boundary detection from motion. Further psychophysical experiments in this area are also suggested. In Chapter 8, some issues for the neurophysiological investigation of discontinuity detection are mentioned briefly.

# 2. BACKGROUND

A particular formulation has been chosen for the computation underlying the measurement of visual motion. Important in this formulation are the input and output representations for the computation. The input representation consists of local measurements of motion along contours in the changing image. These contours correspond to the locations of significant intensity changes, and the local measurements provide the component of motion in the direction perpendicular to the local orientation of the contour. The output representation is an instantaneous two-dimensional velocity field along these contours. In this chapter, some essential background material is presented, that led to this formulation of the motion measurement computation.

## 2.1. Two Processes in Human Vision

The computation of an instantaneous velocity field requires that the movement of elements in the image be roughly continuous. There are alternate representations of visual motion that suggest measurement schemes that are not so restricted. For example, motion can be described by an explicit *correspondence* over time, between elements in the changing image that represent the same physical feature under motion (Ullman 1979a). Motion measurement in this case involves locating identifiable elements in the image, and then matching them over time. In principle, the input for a correspondence scheme may consist of a set of discrete frames, with large spatial separations between corresponding elements. The perception of motion by the human visual system also does not require that a stimulus move continuously across the visual field. With appropriate spatial and temporal presentation parameters, a stimulus presented sequentially can produce the impression of smooth, uninterrupted motion, as in motion pictures (Wertheimer 1912). The visual system can "fill-in" the gaps in the discrete presentation even when the stimuli are separated by up to several degrees of visual angle (Zeeman and Roelofs 1953; Burr and Ross 1982) and by long temporal intervals (up to 400 milliseconds (Neuhaus 1930)). Why, then, have we chosen a formulation of the motion measurement computation that relies on roughly continuous motion?

More recent psychophysical investigation has suggested that in the human visual system, motion may be analyzed by two different systems, termed short range and long range processes by Braddick (1974, 1980). As their names suggest, they are characterized by the range of displacements in space and time over which they operate. It has been proposed that the short range process analyzes continuous motion, or motion that is presented discretely, but with displacement of elements between frames at most $10' - 15'$ of visual arc, and temporal intervals up to 60 — 100 milliseconds. Cinematography usually falls within the limitations of the short range process. The long range process would then analyze motion that is presented discretely, with much larger spatial and temporal separations.

The influence of spatial and temporal parameters on the smoothness of perceived motion was investigated by many researchers (for example, Wertheimer 1912; Korte 1915; Holvand 1935; Neff 1936; Corbin 1942; Graham 1965; Kahneman and Wolman 1970; Kolers 1972; Attneave and Block 1973). A distinction was made between the analysis of short range and long range motion because there exists visual patterns that yield qualitatively different perceptions of motion, depending on the range of spatial and temporal displacements between frames. Examples include the Ternus configuration (Ternus 1926; Pantle and Picciano 1976; Petersik and Pantle 1979), patterns yielding reversed phi (Anstis 1970, 1980; Anstis and Rogers 1975), and dynamic random dot patterns (Anstis 1970; Regan and Spekreijse 1970; Julesz 1971; Lee 1972; Braddick 1973, 1974). A discussion of these phenomena is presented in Chapter 6.

The long range motion phenomena illustrate the ability of the human visual system to derive a correspondence between elements in the changing image, over considerable distances and temporal intervals. Under these conditions, there is no continuous motion of elements across the image to be measured directly. Thus, it is likely that a correspondence computation underlies the long range process. Short range motion is roughly continuous, however; we propose that the computation underlying the short range process may therefore be appropriately formulated as a velocity field computation. This formulation is used for the theoretical analysis of the motion measurement problem presented in the next chapter. We note in Section 3.4, however, that for the analysis of continuous motion, it is possible to integrate aspects of an instantaneous velocity field computation, with those of a correspondence computation.

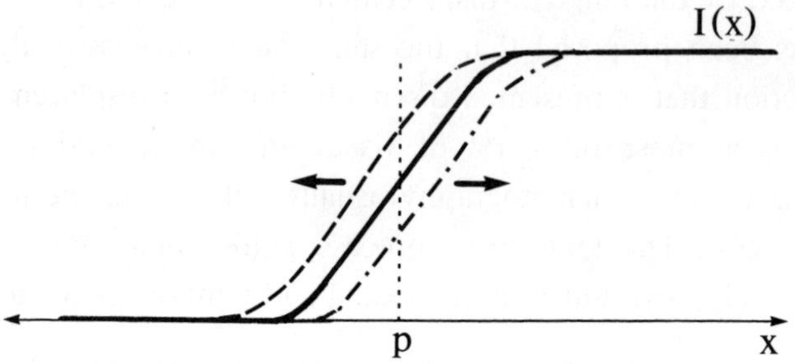


**Figure 2.1. The Gradient Scheme.** The solid curve represents a simple one-dimensional intensity profile. If the curve moves to the left (dashed curve), the intensity $I$ at $p$ increases; for a rightward motion (dotted and dashed curve), $I(p)$ decreases.

## 2.2. The Measurement of Motion Along Contours

In order to detect movement in a changing image, there must be a variation of intensity over space and time; the combination of the two variations can be used to infer the direction and magnitude of velocity. The explicit comparison of spatial and temporal derivatives of intensity forms the basis of a class of motion measurement schemes referred to as *gradient* schemes (Fennema and Thompson 1979; Horn and Schunck 1981; Marr and Ullman 1981). Figure 2.1 shows a one-dimensional example, which illustrates the basic principle underlying the initial detection of motion in gradient schemes. Consider the intensity profile (intensity $I$ as a function of position $x$), indicated by the solid curve in Figure 2.1. At the point $p$, the profile has a positive slope. If the profile moves to the left, indicated by the dashed curve, the intensity value $I$ at $p$ increases; for a rightward motion, indicated by the dotted and dashed curve, $I(p)$ decreases. The sign of the temporal change in $I(p)$ thus signals the direction of motion, and from the magnitude of the spatial and temporal intensity changes, the speed of motion can be determined.[1]

Marr and Ullman (1981) proposed that initial motion measurements in the human visual system are made only at the locations of significant intensity changes, using a gradient scheme. Intensity changes in the image correspond intuitively to

[1]Other examples of motion detection mechanisms, which utilize spatial and temporal intensity changes in a way that differs from gradient schemes, include those suggested by Hassenstein and Reichardt (1956; Reichardt and Poggio 1976) and Barlow and Levick (1965). These latter mechanisms are discussed in Chapters 6 and 8.

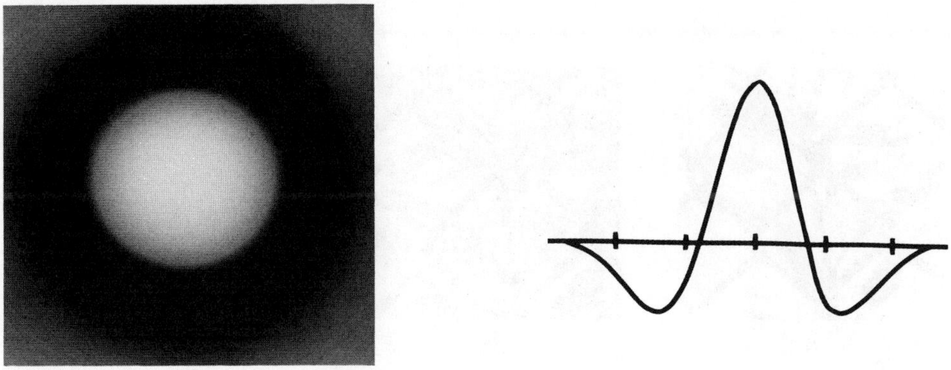

a. b.

**Figure 2.2. The $\nabla^2 G$ operator.** (a) The values of the two-dimensional operator are displayed as grey levels, with white representing the largest positive values, and black indicating extreme negative values. (b) The one-dimensional operator, $D^2 G$.

what we call edges, and are the consequence of changes in some physical property of a surface, such as reflectance, incident illumination, or surface geometry. The utility of edge descriptions for early visual processing has led to an extensive study of methods for detecting and describing intensity changes (reviews of work on edge detection may be found in Davis 1975; Rosenfeld and Kak 1976; Pratt 1978; Brady 1982; Hildreth 1983). A particular representation, called the Primal Sketch, has been suggested by Marr (1976b, 1982). Marr and Hildreth (1980) proposed that for the purpose of detecting intensity changes, a powerful operator for the initial filtering of an image is the Laplacian of a Gaussian, $\nabla^2 G$ (approximated in shape by the difference of two Gaussian functions). The two-dimensional $\nabla^2 G$ operator, and its one-dimensional correlate, $D^2 G$, are shown in Figure 2.2.

The elements in the output of an image convolved with $\nabla^2 G$, which correspond to the locations of significant intensity changes, are the zero-crossings (Marr and Poggio 1979). Figure 2.3 shows an image that has been convolved with a $\nabla^2 G$ operator, and the resulting zero-crossing contours. The size of a $\nabla^2 G$ operator can be defined by the diameter of its positive center; for the operator used to produce the output of Figure 2.3b, this diameter was 9 picture elements. Marr and Hildreth (1980) suggested that in the primate visual system, the convolution of the retinal image with $\nabla^2 G$ (or its approximation, the difference of two Gaussians), is

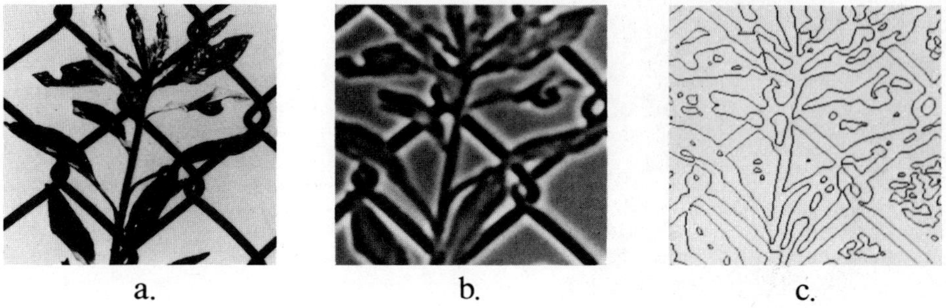

Figure 2.3. **Initial Processing of an Image.** (a) The original image, containing 320 × 320 picture elements. (b) The convolution of the image with a $\nabla^2 G$ operator whose central positive diameter is 9 picture elements. (c) The resulting zero-crossing contours.

represented in the output of the class of retinal ganglion cells referred to as *X-cells* (Rodieck and Stone 1965; Enroth-Cugell and Robson 1966; Cleland, Dubin and Levick 1971; deMonasterio 1978a,b). In primate visual cortex, there is a class of cells, labelled *simple* cells by Hubel and Weisel (1968), that respond preferentially to an edge of a particular orientation and contrast sign in their receptive fields. Many of these cells are also selective for the direction of motion of an edge (Schiller, Finlay and Volman 1976a,b; Poggio, Doty and Talbot 1977). Marr and Hildreth suggested that a class of simple cells may assume a role in the detection of segments of the zero-crossing contours, and proposed a model for the derivation of the spatial selectivity of these cells for zero-crossing contours of a particular orientation and sign of contrast.

Marr and Ullman (1981) extended the model of simple cells proposed by Marr and Hildreth, incorporating a mechanism to account for their selectivity for the direction of motion of a stimulus. A second model, also based on the hypothesis that a class of simple cells may detect moving zero-crossing segments, has been proposed by Poggio (1983). The basic idea underlying Marr and Ullman's model is illustrated in Figure 2.4. Figure 2.4a shows the one-dimensional output of the convolution of a step-edge intensity profile $I$, with the second derivative of a Gaussian, $D^2G$. Figures 2.4b and c illustrate the time derivative of this convolution, $\frac{\partial}{\partial t}(D^2G*I)$, for motion of the profile to the left and right, respectively. At the location of the zero-crossing $Z$,

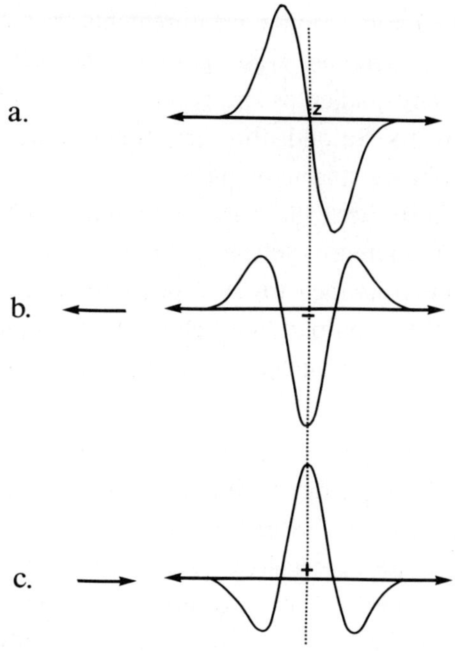


**Figure 2.4. The Marr-Ullman Scheme.** (a) The convolution of a one-dimensional intensity profile, consisting of a single step change of intensity, with the second derivative of a Gaussian, $D^2G$. (b) and (c) The time derivative, $\frac{\partial}{\partial t}(D^2G*I)$, for motion of the profile to the left and right, respectively.

the time derivative is negative for motion to the left, and positive for motion to the right. Thus, the sign of contrast of the zero-crossing can be compared with the sign of the temporal derivative, to compute the direction of motion of the zero-crossing. Marr and Ullman's original model used only the signs of the spatial and temporal derivatives; in principle, a motion detector could be constructed that also combines the rate of change of the convolution output as it crosses zero, with the magnitude of the time derivative, to compute the magnitude of velocity. Further discussion of the initial measurement of motion, from the perspectives of psychophysics and neurophysiology, is presented in Chapters 6 and 8.

From a computational standpoint, restricting the measurement of motion to the location of significant intensity changes, that give rise to zero-crossings in the filtered image, has two advantages over a scheme that bases its initial measurements of motion on variations in the original image intensities, wherever the image intensity gradient is nonzero (such as in the work of Fennema and Thompson (1979) and Horn and Schunck (1981)). First, motion measurements are more reliable where the intensity gradient is steeper. The zero-crossings of $\nabla^2 G * I$ correspond to points in the image at which the intensity gradient is locally maximum, yielding the most reliable local velocity measurements. Second, the zero-crossings are tied more closely to physical features; if they move, it is more likely to be the consequence of movement of an underlying physical surface. There are confounding factors, such as changing illumination, that can cause intensity to change locally. A scheme that infers motion directly from changing intensity is clearly susceptible to incorrectly inferring motion from changes caused by these confounding factors. While the zero-crossing contours may also be influenced by factors such as changing illumination, their positions are generally more stable.

In two dimensions, the spatial and temporal intensity changes alone are not sufficient to determine the local direction and magnitude of velocity, due to the aperture problem. This ambiguity arises from the *local* nature of the initial motion measurement, and is therefore a necessary consequence of any local motion detection mechanism. Motion measurements along contours, for example, provide only the component of velocity in the direction perpendicular to the orientation of the contour. The component of velocity along the contour remains undetected. More formally, the two-dimensional velocity field along a contour may be described by the vector function $\mathbf{V}(s)$, where $s$ denotes arclength. $\mathbf{V}(s)$ can be decomposed into components tangent and perpendicular to the contour, as illustrated in Figure 2.5. $\mathbf{u}^{\top}(s)$ and $\mathbf{u}^{\perp}(s)$ are unit vectors in the directions tangent and perpendicular to the contour, and $v^{\top}(s)$ and $v^{\perp}(s)$ denote the two components:

$$\mathbf{V}(s) = v^{\top}(s)\mathbf{u}^{\top}(s) + v^{\perp}(s)\mathbf{u}^{\perp}(s). \tag{1}$$

The component $v^{\perp}(s)$, and direction vectors $\mathbf{u}^{\top}(s)$ and $\mathbf{u}^{\perp}(s)$, are given directly by the initial measurements from the changing image. The component $v^{\top}(s)$ is not, and must be recovered, to compute $\mathbf{V}(s)$. Intuitively, the set of measurements given by $v^{\perp}(s)$ over an extended contour should provide considerable constraint on the motion of the contour. Additional constraint is still required, however, to determine this motion uniquely. Thus the computation of $\mathbf{V}(s)$ requires the integration of the constraints provided by $v^{\perp}(s)$ along the contour, together with additional constraints necessary to compute $\mathbf{V}(s)$ uniquely.

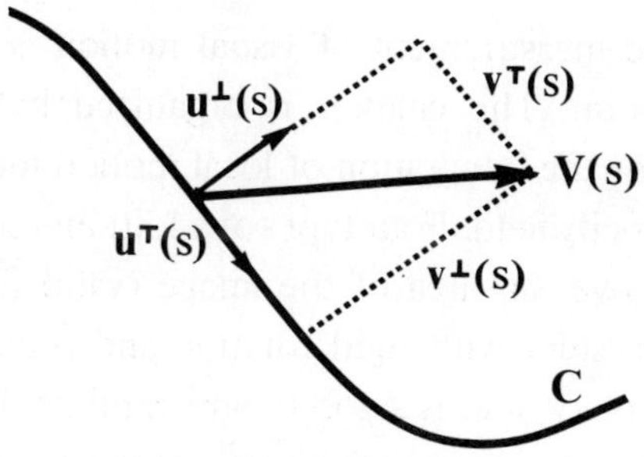


**Figure 2.5. Decomposition of Velocity.** The local velocity vector $V(s)$ is decomposed into components perpendicular and tangent to the curve.

To summarize, Marr and Ullman (1981) proposed that the measurement of motion by the short range process in human vision is focused along significant intensity changes in the image, which give rise to zero-crossings in the output of an image filtered with a $\nabla^2 G$ operator. The initial motion measurements provide only one component of velocity, and must be combined to compute the full two-dimensional velocity field. For much of the theoretical discussion of the motion measurement problem presented in the next chapter, we assume an input representation for the velocity field computation that consists of image contours, and some measure of the component of velocity in the direction perpendicular to the contours. The theoretical results do not rely on the particular scheme proposed by Marr and Ullman for obtaining this representation. In addition, many of the basic ideas may be used in other approaches to the motion measurement problem; for example, in a motion correspondence computation, or in a velocity field computation that is not based on motion measurements along contours.

# 3. COMPUTATION OF THE VELOCITY FIELD

In this chapter, the measurement of visual motion is discussed, strictly from a computational viewpoint. The chapter is organized by the type of additional constraint that is used for the integration of local motion measurements to compute the two-dimensional velocity field. Four types of additional constraint are considered: (1) velocity is constant over an area of the image (valid for pure translation); (2) the velocity field is consistent with rigid rotation and translation of objects in the image plane; (3) the velocity field is smooth, and exhibits the least variation among the set of velocity fields consistent with the changing image; and (4) the velocity field is constant over small time intervals.

The discussion of the constant velocity constraint, in Section 3.1, is primarily a review of previous work on the motion measurement computation. This assumption underlies most schemes that have been proposed, both for computer and biological vision systems. The class of rigid motions in the image plane has been addressed in correspondence schemes for motion measurement. In Section 3.2, a scheme is presented for analyzing the instantaneous velocity field for this class of motions. Methods that are based on the assumptions of constant velocity, or rigid motion in the image plane, are not sufficient for analyzing the two-dimensional motion that arises from the projection of arbitrary three-dimensional surfaces undergoing general motion in space. They may be useful, however, for the rough analysis of movement in the periphery in the case of a biological system, or for restricted applications of computer vision systems.

The constraint on smoothness of the velocity field, discussed in Section 3.3, is motivated by the work of Horn and Schunck (1981) on the optical flow computation, and allows the analysis of more general classes of motion. We apply the smoothness constraint along contours in the image, and show that this constraint may be formulated precisely in a way that guarantees a unique velocity field solution. Some theoretical results are also presented, concerning the physical plausibility of the velocity field obtained from a computation that embodies this constraint. The final section of this chapter discusses an integrated approach to the motion measurement problem, that combines instantaneous motion measurements with a correspondence computation.

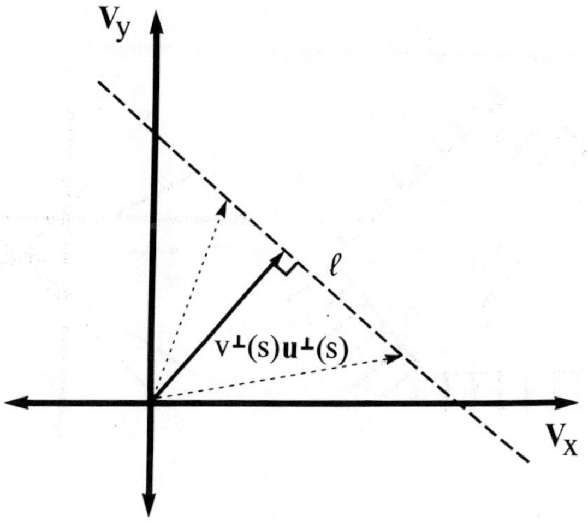


**Figure 3.1. Velocity Constraints in Velocity Space.** The vector $v^{\perp}(s)\mathbf{u}^{\perp}(s)$ is the perpendicular component of velocity at a point $p$ on the image contour. The velocity vector at $p$ must project to the line $l$; examples are shown with dotted lines.

### 3.1. The Constant Velocity Constraint

Much of the previous work in motion analysis has addressed the case of pure translation of objects in the image plane. Most gradient schemes, for example, assume that velocity is constant over an area of the image (Limb and Murphy 1975; Fennema and Thompson 1979; Marr and Ullman 1981; Thompson and Barnard 1981; Schunck 1983). For gradient schemes, the constraint on velocity imposed by a single measurement of $v^{\perp}(s)$ can be illustrated graphically in *velocity space*, a space in which the $x$ and $y$ axes represent the $x$ and $y$ components of velocity, which we denote by $\mathbf{V}_x$ and $\mathbf{V}_y$. This is shown in Figure 3.1. When mapped to velocity space, the velocity vector at a point on the contour must terminate along the line $l$, which is perpendicular to the vector $v^{\perp}(s)\mathbf{u}^{\perp}(s)$; examples are shown by the dotted arrows. For the case of uniform translation, the lines of constraint formed by the measurements of $v^{\perp}(s)$ along a contour intersect at a single point in velocity space. Figure 3.2a shows a simple polygon translating in the image plane with velocity $\mathbf{V}_o$. The arrows indicate the perpendicular components of velocity, for points on each side of the polygon. The constraints on the true velocity of the contour, imposed by $v^{\perp}(s)$, are given by the lines in velocity space shown in Figure 3.2b. The lines intersect at the endpoint of the vector $\mathbf{V}_o$, shown as a dotted line.

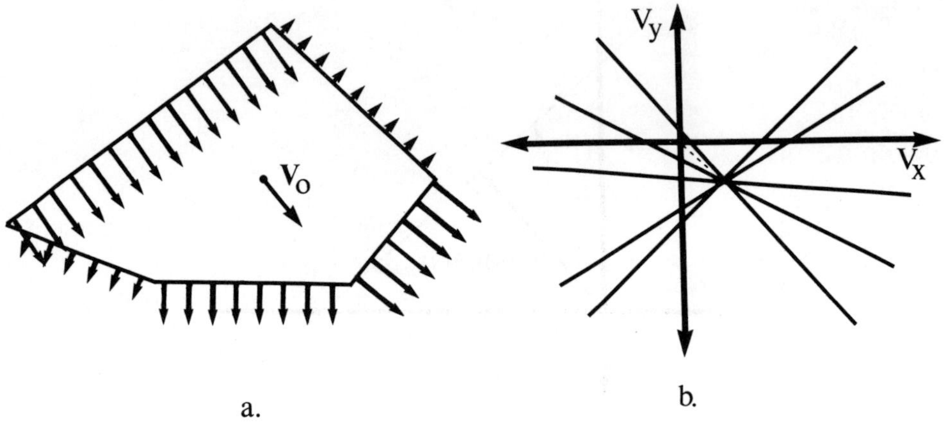


**Figure 3.2. Rigid Motion in the Image Plane.** (a) A simple polygon translating in the image plane with velocity $V_o$; the arrows indicate the perpendicular components of velocity. (b) The lines of constraint in velocity space intersect at a single point.

Several motion measurement schemes have been developed, which make use of the intersection of multiple motion constraints in velocity space (for example, Fennema and Thompson 1979; Marr and Ullman 1981; Thompson and Barnard 1981; Adelson and Movshon 1982; Schunck 1983). Adelson and Movshon (1982) suggest the direct use of the intersection of lines of constraint, illustrated in Figure 3.2b. Thompson and Barnard (1981) describe a linear systems optimization technique for computing this intersection, that allows for error in the initial motion measurements. In an earlier gradient scheme, Fennema and Thompson (1979) formulated the local motion constraints in polar coordinates rather than cartesian; a single measurement then constrained local velocity to lie along a cosine curve in velocity space, rather than a straight line. A modified Hough transform was used to compute the intersection of multiple motion constraints. A maximum likelihood estimation method for computing the velocity field from initial gradient measurements was developed by Schunck (1983). Marr and Ullman (1981; Batali and Ullman 1979) proposed a scheme for computing the motion of zero-crossings, in which each local measurement restricts the true direction of velocity of a patch to lie within a 180° range of directions to one side of a segment of the local zero-crossing contour. A set of measurements taken at different orientations along a contour then further restrict the allowable velocity directions, until a single direction is obtained, which is consistent with all the local measurements.

Motion measurement schemes based on the cross-correlation of intensity, used both in computer vision (Smith and Phillips 1972; Leese, Novak and Taylor 1970; Lillestrand 1972; Wolferts 1974) and in modelling biological vision systems (Reichardt 1961; Anstis 1970, 1980; Bell and Lappin 1973; Lappin and Bell 1976; Pantle and Picciano 1976; Petersik, Hicks and Pantle 1978), also rely on pure translation. Correspondence schemes, such as those presented by Lawton (1983), Potter (1975, 1977), and Mutch and Thompson (1982), assume local translation of features in the image, although there is variation in the particular features that are matched between frames. In Lawton's work, the matching elements are *distinctive* points, usually points of high curvature, along zero-crossing contours. Mutch and Thompson use extrema in the Laplacian of a blurred image, and Potter uses primitive edge elements (grey level discontinuities) and templates that characterize geometric relationships between edges. In addition to the above work, schemes for analyzing rigid rotation and translation in the image plane, discussed in the next section, may be used for the restricted case of pure translation.

In many natural dynamic scenes, particularly those arising from motion of the observer, the velocity field can be approximated locally by pure translation. Most of the above schemes involve local operations for computing displacements or velocities, and can be effective for analyzing this type of motion. In more general situations, for example, when a nearby surface rotates in depth and deforms, the assumption of local translation is no longer sufficient.

### 3.2. Rigid Motion in the Image Plane

In this section, we consider the restricted class of rigid motions in the image plane, and present some results that lead to a simple algorithm for computing the local velocity field. These results are also extended to the case in which a curve undergoes a uniform expansion or contraction in the image, in addition to a rotation and translation. Throughout the remainder of this chapter, the terms *contour* and *curve* will be used interchangeably.

Suppose that a rigid curve undergoes general motion in space. Its instantaneous motion may be described as the combination of (1) a rotation with angular velocity

$\omega$ about a single axis in space, which we denote by the unit vector $\mathbf{n} = [n_x, n_y, n_z]^T$ ($T$ denotes the transpose of a vector), and (2) a translation, which we denote by the vector $\mathbf{t} = [t_x, t_y, t_z]^T$. Let the curve be given by $C = (x(s), y(s), z(s))$, where $s$ denotes arclength. The location of a point on the curve may be given by the position vector $\mathbf{r} = [x(s), y(s), z(s)]^T$. If it is assumed that the axis of rotation passes through the origin of the coordinate system, the velocity of a point given by the position vector $\mathbf{r}$ is equal to the cross product of $\mathbf{r}$ with the vector $\omega\mathbf{n}$. If we then let the optical axis lie along the $z$-axis, and let the projection of the curve onto the image plane (the $(x, y)$ plane) be orthographic, then the two-dimensional velocity field $\mathbf{V}(s)$ along the contour is given by:

$$\mathbf{V}(s) = \mathbf{M}(\mathbf{r} \times \omega\mathbf{n} + \mathbf{t}) = \omega z(s)\begin{bmatrix} n_y \\ -n_x \end{bmatrix} + \omega n_z\begin{bmatrix} -y(s) \\ x(s) \end{bmatrix} + \begin{bmatrix} t_x \\ t_y \end{bmatrix}. \qquad (2)$$

$\mathbf{M}$ denotes the matrix that performs the orthographic projection. The first term in the resulting expression describes the component of the velocity field due to rotation in depth about an axis parallel to the image plane (the axis $\mathbf{n} = [n_x, n_y, 0]^T$); the second term is the component due to motion in the image plane (rotation about the axis $\mathbf{n} = [0, 0, n_z]^T$), and the third term is the translation component.

Consider the restricted case of rigid motion in the image plane; the velocity field now corresponds to the combination of a translation, and rotation about the axis $\mathbf{n} = [0, 0, 1]^T$. Thus, $\mathbf{V}(s)$ is given by:

$$\mathbf{V}(s) = \omega\begin{bmatrix} -y(s) \\ x(s) \end{bmatrix} + \begin{bmatrix} t_x \\ t_y \end{bmatrix}. \qquad (3)$$

$\mathbf{V}(s)$ is simply a translation, rotation and scaling of the image curve $(x(s), y(s))$, as illustrated in Figure 3.3. In Figure 3.3a, the curve $\mathbf{C}_1$ undergoes a small rotation and translation in the image plane to yield the curve $\mathbf{C}_2$. The arrows indicate a sampling of the local velocity vectors along the curve. In Figure 3.3b, these velocity vectors have been translated to a common origin in velocity space. The curve in velocity space has the same shape as the image curve $\mathbf{C}_1$; its size is proportional to angular velocity $\omega$, and it is rotated 90° with respect to $\mathbf{C}_1$ (this relationship is also used in kinematics (Hartenberg and Denavit 1964)). The additional translation of the curve in the image yields the same translation of the curve in velocity space. In the case of pure translation, the image of the velocity field in velocity space

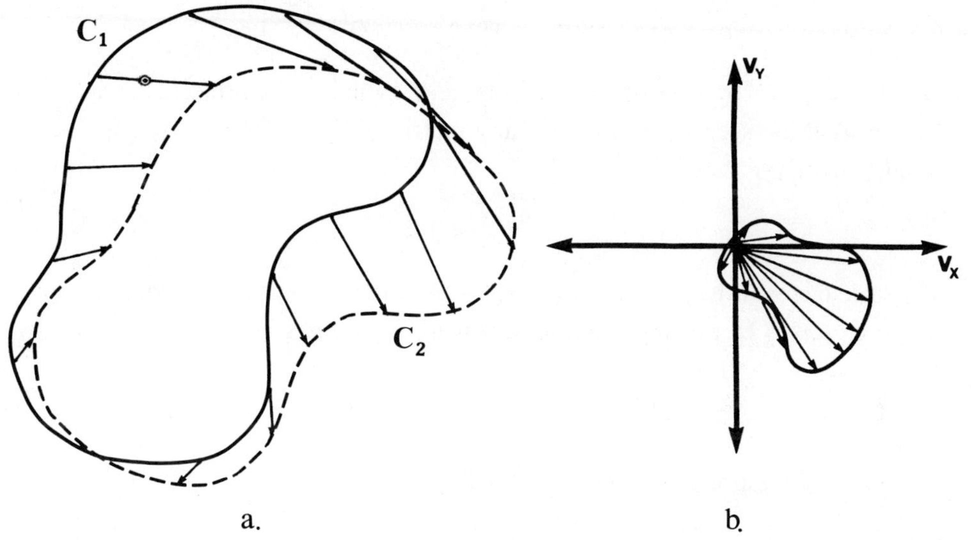


**Figure 3.3. Velocity Field for Rigid Motion in the Image Plane.** (a) The image curve $\mathbf{C}_1$ undergoes a small rotation and translation in the image plane to yield the curve $\mathbf{C}_2$. The arrows represent local velocity vectors. (b) The velocity vectors drawn in velocity space.

degenerates to a single vector. In general, the explicit use of the velocity space aids in the visualization of properties of the motion of curves, and provides a tool for establishing theoretical properties of the velocity field.

For the simple case of rigid motion in the image plane, this relationship between the shape of the curve and the velocity field is not restricted to continuous motion of the curve. For discrete motion of a curve, we use the term *displacement field* for the set of vectors that describe the discrete displacement of points on the curve, and denote this field by $\mathbf{D}(s)$. Let $\mathbf{C}_1 = (x_1(s), y_1(s))$ denote the initial curve, and $\mathbf{C}_2 = (x_2(s), y_2(s))$ denote the new curve that results from a rotation and translation of $\mathbf{C}_1$. Further, let the discrete translation be given by the vector $\mathbf{t} = [t_x, t_y]^T$, and let $\sigma$ be the discrete angular rotation of the curve in the image plane. The curve $\mathbf{C}_2$ is then given by the following:

$$\begin{bmatrix} x_2(s) \\ y_2(s) \end{bmatrix} = \begin{bmatrix} \cos\sigma & -\sin\sigma \\ \sin\sigma & \cos\sigma \end{bmatrix} \begin{bmatrix} x_1(s) \\ y_1(s) \end{bmatrix} + \begin{bmatrix} t_x \\ t_y \end{bmatrix}. \tag{4}$$

To obtain $\mathbf{D}(s)$, the initial position of the curve is subtracted, giving the following:

$$\mathbf{D}(s) = \begin{bmatrix} \cos\sigma - 1 & -\sin\sigma \\ \sin\sigma & \cos\sigma - 1 \end{bmatrix} \begin{bmatrix} x_1(s) \\ y_1(s) \end{bmatrix} + \begin{bmatrix} t_x \\ t_y \end{bmatrix}. \quad (5)$$

$\mathbf{D}(s)$ is also given by a translation, rotation and scaling of the projected image curve $\mathbf{C}_1$. This can be seen by noting that any matrix of the following form is a scaled rotation matrix:

$$\begin{bmatrix} a & -b \\ b & a \end{bmatrix}. \quad (6)$$

The scale factor, $k$, is given by: $k = \sqrt{a^2 + b^2}$, and the angle of rotation, $\alpha$, is given by: $\alpha = tan^{-1}(\frac{b}{a})$. For the displacement field, $\mathbf{D}(s)$, the scale factor $k$ is given by:

$$k = \sqrt{(\cos\sigma - 1)^2 + \sin^2\sigma} = \sqrt{2(1 - \cos\sigma)}. \quad (7)$$

The angle of rotation $\alpha$ between $\mathbf{C}_1$ and $\mathbf{D}(s)$ is given by:

$$\tan\alpha = \frac{\sin\sigma}{\cos\sigma - 1} = -\cot(\frac{\sigma}{2}). \quad (8)$$

For small $\sigma$, $k \approx \sigma$ and $\alpha \approx \pm 90°$. As before, an additional translation component simply translates the curve in velocity space.

For rigid motion in the image plane, a simple geometric construction can be used to compute the velocity field. If the true direction of velocity is known at two points on the contour, the direction of velocity everywhere can be computed as follows: (1) construct the lines perpendicular to the direction of velocity at the two known points, (2) compute the intersection of these two lines, (3) from every point $p_i$ along the contour, construct the line to the intersection point; the true direction of velocity is perpendicular to this line. In Figure 3.4, the direction at $p_2$ is derived, given known directions at $p_1$ and $p_3$.

This geometric construction is simply locating the point about which the motion can be described as pure rotation. For pure translation, the two lines, from points of known direction of velocity, are parallel, so the direction of motion everywhere is equal to the direction of motion of the known points. Certainly, if both the direction and magnitude of velocity were known at two points on the contour, the global motion parameters, and hence direction and magnitude of velocity everywhere, could be computed. From the direction of velocity alone at two points on the

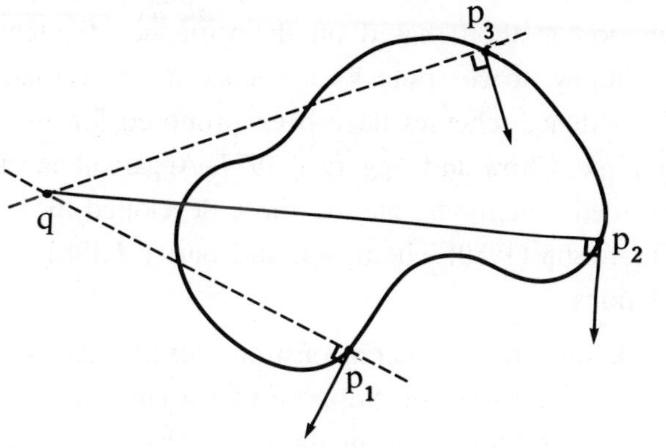


**Figure 3.4. Constructing the Velocity Field.** The direction of velocity is known at points $p_1$ and $p_3$. Using the construction scheme described in the text, the direction of velocity at the point $p_2$ is computed.

contour, however, the direction of velocity everywhere can be obtained. If, in addition, the perpendicular components of velocity are known, then both direction and magnitude of velocity can be computed along the contour.

There are at least two sources for points of known velocity direction in the image. First, identifiable features, such as terminations, may be tracked in two dimensions. The tracking of identifiable features along zero-crossing contours forms the basis of the motion measurement scheme proposed by Lawton (1983). As a second source of known directions, points at which the perpendicular component of velocity is zero are constrained to move along the direction of the tangent to the curve. For the case of a smooth, closed curve moving rigidly in the image plane, there exists at least two points on the curve for which the perpendicular component of velocity is zero (for proof, see Appendix I). Suppose the velocity field computation is focused at the locations of the zero-crossings derived from the convolution of the image with the $\nabla^2 G$ operator. These zero-crossing contours are generally closed (except at image boundaries), so there is usually sufficient constraint from the image to compute the velocity field, for this simple class of motions.

Other motion measurement schemes have been proposed for analyzing rigid motion in the image plane. Davis, Wu and Sun (1982) present an iterative gradient scheme for combining motion constraints along contours. The scheme starts from points of known velocity, and propagates motion information along contours, using the perpendicular components of velocity. Nagel (1982) presents a maximum likelihood estimation method, based on the variance of intensity values within a small image window, to compute local displacement vectors. In limited image domains, correspondence schemes have been proposed for analyzing this class of motions (for example, Chow and Aggarwal 1977; Aggarwal and Duda 1975). More general correspondence methods, such as those developed by Ullman (1979a) and Barnard and Thompson (1980; Thompson and Barnard 1981) also allow rotations as well as translations.

Before considering a more general constraint for the velocity field computation, we mention a second special case. Suppose that a curve **C** were also to undergo a uniform expansion or contraction in the image, by a factor $\rho$. This could arise physically from a small translation of the curve along the optical axis, under perspective projection. If we consider the case of a discrete motion, and again let the angle of rotation be $\sigma$, the displacement field is given by:

$$\mathbf{D}(s) = \begin{bmatrix} \rho\cos\sigma - 1 & -\rho\sin\sigma \\ \rho\sin\sigma & \rho\cos\sigma - 1 \end{bmatrix} \begin{bmatrix} x(s) \\ y(s) \end{bmatrix}. \tag{9}$$

The curve in velocity space is again a simple rotation and scaling of the image curve; the scale factor $k$ relating the sizes of **C** and $\mathbf{D}(s)$ is now given by:

$$k = \rho\sqrt{2(1 - \cos\sigma)}. \tag{10}$$

The rotation $\alpha$ between **C** and $\mathbf{D}(s)$ is given by:

$$\tan\alpha = \frac{\rho\sin\sigma}{\rho\cos\sigma - 1}. \tag{11}$$

Referring to Figure 3.5a, if the displacement vectors are known at two points on the curve, which we denote by $\mathbf{D}(s_1)$ and $\mathbf{D}(s_2)$, then the scale factor $k$ may be computed as follows:

$$k = \frac{|\mathbf{D}(s_2) - \mathbf{D}(s_1)|}{|\mathbf{r}(s_2) - \mathbf{r}(s_1)|} \tag{12}$$

where $\mathbf{r}(s_1)$ and $\mathbf{r}(s_2)$ are the position vectors for the two points. $\alpha$ is the angular difference between the vectors $[\mathbf{r}(s_2) - \mathbf{r}(s_1)]$ and $[\mathbf{D}(s_2) - \mathbf{D}(s_1)]$. Given a rotation of $\alpha$ (rather than $90^\circ$ as before), the geometric construction of Figure 3.4 can be

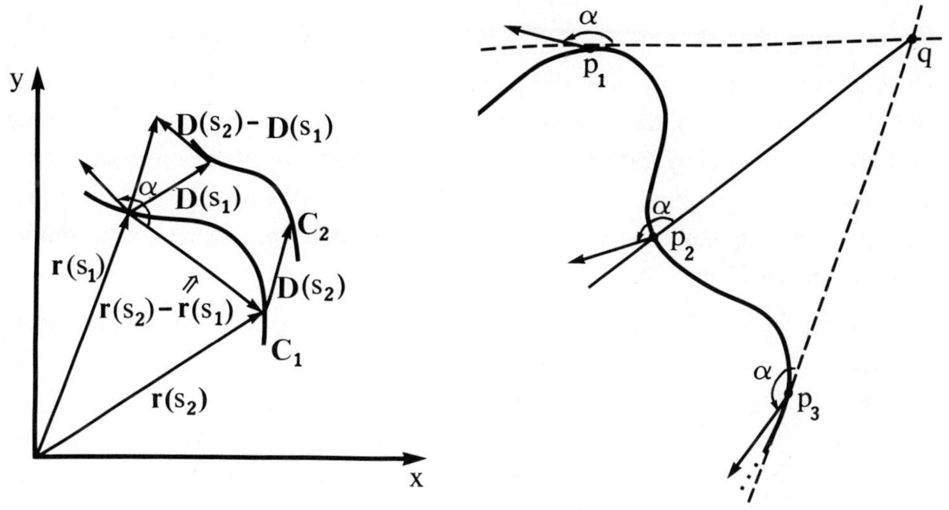


a. b.

**Figure 3.5. Adding a Uniform Scaling.** (a) The curve $\mathbf{C}_1$ undergoes a small rotation, translation and scaling to yield the curve $\mathbf{C}_2$. The position vectors $\mathbf{r}(s_1)$ and $\mathbf{r}(s_2)$ and displacement vectors $\mathbf{D}(s_1)$ and $\mathbf{D}(s_2)$ can be used to compute the scale factor $k$. (b) Given the angle $\alpha$, the construction scheme is adapted to derive the direction of velocity at a point $p_2$, given known directions at $p_1$ and $p_3$.

adapted to compute the local velocity directions, as shown in Figure 3.5b. The direction of velocity at a point $p$ on the curve now forms the angle $\alpha$ with the line connecting $p$ and the intersection point $q$. In Figure 3.5b, the direction of velocity is computed for the point $p_2$, given known directions at $p_1$ and $p_3$. As before, the magnitudes of the perpendicular components of velocity can then be used to compute the magnitude of velocity everywhere.

The above methods for computing the velocity field, developed for the restricted class of rigid motions in the image plane, are not sufficient for general motion analysis; however, they may be useful for the initial detection and rough measurement of motion in the periphery, the analysis of motion during smooth pursuit eye movements, or the recovery of observer motion from optical flow (Prazdny 1980; Longuet-Higgins and Prazdny 1981). In computer vision, there are restricted applications for these techniques, such as the tracking of objects, or computation of camera motion (Bruss and Horn 1983; Nagel 1981; Lawton 1983).

### 3.3. The Smoothness Constraint

In this section, a different type of constraint on the velocity field is derived, which allows for the analysis of the projected motion of three-dimensional objects that move freely in space, and deform over time. The specific analysis assumes an input representation for the velocity field computation that consists of image contours, and the perpendicular components of velocity along these contours. The general constraint, however, also may be used in other approaches to the motion measurement problem.

The expression (2) in Section 3.2 related $\mathbf{V}(s)$ to the global motion parameters $\omega$ and $\mathbf{n}$, and the shape of the curve $\mathbf{C} = (x(s), y(s), z(s))$, for the case in which the curve undergoes rigid motion in space. The relationship between $\mathbf{V}(s)$ and $\mathbf{C}$ is quite simple. If the projected two-dimensional velocity vectors, given by $\mathbf{V}(s)$, are mapped to a common origin in velocity space, their endpoints trace out a scaled, $90°$ rotation of the projected image curve, with an additional distortion along one direction. This distortion is directed perpendicular to the axis $\mathbf{n} = [n_x, n_y, 0]^T$, and is scaled by the $z$ component of the curve, $z(s)$. This relationship implies, for example, that a smooth curve in motion usually generates a smoothly varying velocity field. The real world consists predominantly of solid objects, whose surfaces are generally smooth compared with their distance from the viewer. Thus, intuitively, we seek a velocity field that is consistent with the changing image, and which varies smoothly. Unfortunately, there is an infinity of possible velocity fields that satisfy these two properties. Horn and Schunck (1981), in their work on the optical flow computation, suggest that a single solution may be obtained by finding the velocity field that varies the least, among the set of possible velocity fields consistent with the changing image. In this section, we show how this constraint can be formulated precisely, in a way that yields a velocity field solution that is mathematically unique and physically plausible.

#### 3.3.1. Measuring Variation in Velocity

To find the velocity field that varies the least, some means of measuring the variation in velocity along a contour is required. This can be accomplished in various ways. For example, we could measure the change in direction of velocity as we trace along the contour. Total variation of the velocity field could be defined as the total change in direction over the entire contour. A second possibility is to measure the change in magnitude of velocity along the contour. Third, the change in the full two-dimensional velocity vector could be measured, incorporating both the direction and magnitude of velocity. Other measures are also possible. The goal of the computation is to find a velocity field that is consistent with the changing image, and minimizes one of these measures of variation in velocity.

In order to define the variation in the velocity field more formally, first recall the decomposition of velocity into components tangent and perpendicular to the curve:

$$\mathbf{V}(s) = v^{\top}(s)\mathbf{u}^{\top}(s) + v^{\perp}(s)\mathbf{u}^{\perp}(s). \tag{1}$$

$\mathbf{u}^{\top}(s)$, $\mathbf{u}^{\perp}(s)$ and $v^{\perp}(s)$ can be measured directly from the changing image. $v^{\top}(s)$ is unknown, and must be recovered in order to compute the velocity field $\mathbf{V}(s)$. Aside from knowing $v^{\perp}(s)$ everywhere along the curve, there may also be points at which the direction and magnitude of velocity, and hence both $v^{\perp}(s)$ and $v^{\top}(s)$, are known. In addition, the direction of velocity alone may be known at points on the curve, for example, where $v^{\perp}(s_i) = 0$ (see Section 3.2).

A more formal means for measuring variation in the velocity field can now be considered. Mathematically, this can be accomplished by defining a functional $\Theta$, which maps the space of all possible vector fields (along the contour), $V$, into the real numbers: $\Theta : V \mapsto \Re$. This functional should be such that the smaller the variation in the velocity field, the smaller the real number assigned to it. Two candidate velocity fields may then be compared, by comparing their corresponding real numbers. This raises the question of what functional should be used to measure the variation in a velocity field. In the remainder of this section, we evaluate a set of possible functionals, based on the three measures of variation that were previously mentioned informally: (1) variation in $\mathbf{V}(s)$, (2) variation in the direction of velocity, and (3) variation in the magnitude of velocity, all with respect to the curve.

**1. Variation in $\mathbf{V}(s)$**

The local variation in $\mathbf{V}(s)$ with respect to the curve is given by $\frac{\partial \mathbf{V}}{\partial s}$. A scalar measure may be obtained by taking its magnitude: $\left|\frac{\partial \mathbf{V}}{\partial s}\right|$. In Figure 3.6a two nearby velocity vectors along the image curve **C** are displayed. The vectors are translated to a common origin in velocity space in Figure 3.6b, where the vector $\frac{\partial \mathbf{V}}{\partial s}$ is shown with a dotted arrow. A measure of the total variation in velocity along a curve may

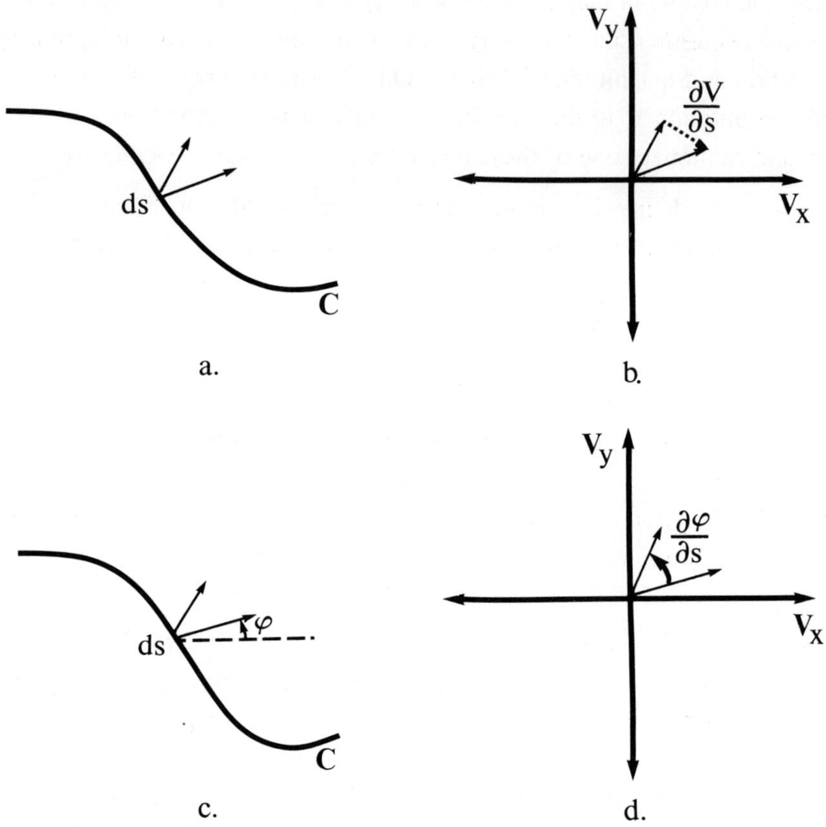


**Figure 3.6. Measuring Variation in Velocity.** (a) The velocity vectors $\mathbf{V}(s)$ are displayed at two points on the image curve. (b) The velocity vectors drawn in velocity space, where $\frac{\partial \mathbf{V}}{\partial s}$ is indicated by the dotted arrow. (c) The direction of velocity for points on the contour is represented by the angle $\varphi$. (d) The velocity vectors of (c) are drawn in velocity space, where $\frac{\partial \varphi}{\partial s}$ is shown.

be derived by integrating this local measure, suggesting a functional such as:

$$\Theta(\mathbf{V}) = \int \left| \frac{\partial \mathbf{V}}{\partial s} \right| ds. \tag{13}$$

Variations on this functional may also be considered, involving higher order derivatives, or higher powers, such as:

$$\Theta(\mathbf{V}) = \int \left| \frac{\partial^2 \mathbf{V}}{\partial s^2} \right| ds \quad \text{or} \quad \Theta(\mathbf{V}) = \int \left| \frac{\partial \mathbf{V}}{\partial s} \right|^2 ds. \tag{14}$$

**2. Variation in Direction**

Let the direction of velocity be given by the angle $\varphi$, measured in the counterclockwise direction from the horizontal, as shown in Figure 3.6c. In Figure 3.6d, the local change in direction, for two nearby velocity vectors along the image curve, given by $\frac{\partial \varphi}{\partial s}$, is shown in velocity space. Total variation in direction along the curve can again be obtained by integrating this local measure, leading to functionals such as the following:

$$\Theta(\mathbf{V}) = \int \left| \frac{\partial \varphi}{\partial s} \right| ds \tag{15}$$

or variations involving higher order derivatives, or higher powers.

**3. Variation in Magnitude**

Finally, the total change in magnitude of velocity alone could be measured, using functionals such as:

$$\Theta(\mathbf{V}) = \int \frac{\partial |\mathbf{V}|}{\partial s} ds. \tag{16}$$

Again, variations on this measure could also be considered.

The functional that is used to measure variation may also incorporate a measure of the velocity field itself, rather than strictly utilizing changes in the velocity field along the curve. For example, we could incorporate a term that is a function of $|\mathbf{V}|$. This might be useful if we sought a velocity field that also exhibits the least total motion. In addition, the functional could become arbitrarily complex in its combination of $|\frac{\partial \mathbf{V}}{\partial s}|$, $|\frac{\partial \varphi}{\partial s}|$, $\frac{\partial |\mathbf{V}|}{\partial s}$, or higher order derivatives.

Given that there are many possible measures of variation, what criteria can be used to choose a single measure? First, from a mathematical point of view, there should exist a unique velocity field that minimizes the particular measure of variation; this requirement imposes a set of mathematical constraints on the functional. Second, the velocity field computation should yield solutions that are physically plausible. Third, if we suggest that such a constraint underlies the motion computation in the human visual system, the minimization of this measure of variation should yield a velocity field that is consistent with human motion perception. These three criteria are important for the evaluation of any additional assumption for the motion measurement computation.

### 3.3.2. Mathematical Uniqueness of the Velocity Field

An examination of these measures of variation from a mathematical point of view suggests that a measure incorporating the change in the full two-dimensional velocity vector is necessary for the velocity field computation. The use of functionals that incorporate only a measure of direction or magnitude of velocity, for example, does not in general lead to a unique velocity field solution. It can be shown, however, that given a simple condition on the constraints that are derived from the image, there exists a unique velocity field that satisfies these constraints, and minimizes the particular measure of variation given by: $\int |\frac{\partial \mathbf{V}}{\partial s}|^2 ds$. To obtain this result, we take advantage of the analysis used by Grimson (1981) for evaluating possible functionals for performing surface interpolation from sparse stereo data. The basic mathematical question is, what conditions on the form of the functional, and the structure of the space of velocity fields, are needed to guarantee the existence of a unique solution? These conditions are captured by the following theorem from functional analysis (see also (Rudin 1973)):

> <u>**Theorem**</u>: *Suppose there exists a complete semi-norm $\Theta$ on a space of functions $H$, and that $\Theta$ satisfies the parallelogram law. Then, every nonempty closed convex set $E \subset H$ contains a unique element $v$ of minimal norm, up to an element of the null space. Thus, the family of minimal functions is*
>
> $$\{v + s \mid s \in S\}$$
>
> *where*
>
> $$S = \{w \mid v + w \in E\} \cap \mathcal{N}$$
>
> *and $\mathcal{N}$ is the null space of the functional*
>
> $$\mathcal{N} = \{u \mid \Theta(u) = 0\}.$$

In Appendix IIa, we first show that the functional $\Theta = \{\int |\frac{\partial \mathbf{V}}{\partial s}|^2 ds\}^{\frac{1}{2}}$ is a complete semi-norm that satisfies the parallelogram law. Second, we show that the space of all possible velocity fields that satisfy the constraints derived from the image, is convex. It then follows from the above theorem that this space contains a unique element of minimal norm, up to possibly an element of the null space. The measure of variation is non-negative, so that minimizing $\{\int |\frac{\partial \mathbf{V}}{\partial s}|^2 ds\}^{\frac{1}{2}}$ is equivalent to minimizing $\int |\frac{\partial \mathbf{V}}{\partial s}|^2 ds$.

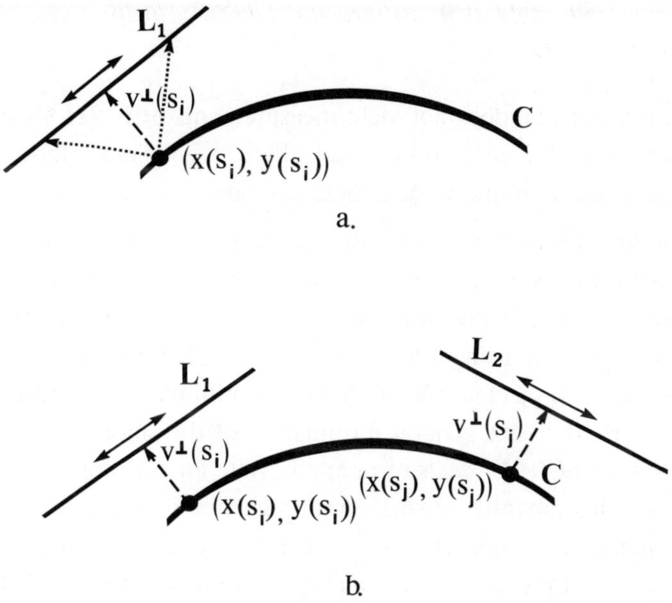


**Figure 3.7. Uniqueness of the Velocity Field.** (a) The constraint imposed by a single measurement of $v^{\perp}(s)$ on the contour **C**. Only a uniform translation along the line $\mathbf{L}_1$ can be added to the velocity field. (b) The constraint imposed by two measurements of $v^{\perp}(s)$ on **C**.

The null space in this case is the set of constant velocity fields, because the condition that $\int|\frac{\partial \mathbf{V}}{\partial s}|^2 ds = 0$ implies that $\left|\frac{\partial \mathbf{V}}{\partial s}\right| = 0$ everywhere, which implies that $\mathbf{V}(s)$ is constant. Suppose there is a point $(x(s_i), y(s_i))$ on the curve, where $v^{\perp}(s_i)$ is known. This measurement constrains the velocity $\mathbf{V}(s_i)$ to lie along the line $\mathbf{L}_1$ parallel to the tangent of the curve at this point, as shown in Figure 3.7a. Suppose there is a velocity field that is consistent with $v^{\perp}(s_i)$. A uniform translation component can then be added only along the direction of $\mathbf{L}_1$, in order to obtain another velocity field that is still consistent with this local measure. If a second measurement $v^{\perp}(s_j)$ is known at a point $(x(s_j), y(s_j))$, for which the direction of the tangent is different (see Figure 3.7b), then a uniform translation component can be added only along this second direction, in order to obtain a velocity field that is still consistent with $v^{\perp}(s_j)$. A uniform translation cannot be added to the entire velocity field, which is consistent with both local measurements. Thus, we conclude the following:

*If $v^{\perp}(s)$ is known along a contour, and there exists at least two points at which the local orientation of the contour is different, then there exists a unique velocity field that satisfies the known velocity constraints and minimizes* $\int |\frac{\partial \mathbf{V}}{\partial s}|^2 ds$.

An extended straight line does not yield measurements of $v^{\perp}(s)$ for points at two different orientations, but in all other cases, there is sufficient information along a contour to guarantee a unique velocity field solution.

The smoothness constraint can be used to compute a projected two-dimensional velocity field for any three-dimensional surface, whether rigid or nonrigid, undergoing general motion in space. If the variation in the full velocity vector along contours in the image is measured, using a functional such as $\int|\frac{\partial \mathbf{V}}{\partial s}|^2 ds$, it is guaranteed that there exists a unique projected velocity field that minimizes this variation. While it is not yet clear whether this general formulation of the smoothness constraint, or the particular measure $\int|\frac{\partial \mathbf{V}}{\partial s}|^2 ds$, is most appropriate for the motion computation, it is important that this measure satisfies certain essential mathematical requirements, that the other measures do not. The use of a functional incorporating only a measure of velocity direction, for example, which attempts to make the local velocity vectors as parallel as possible, does not lead to a unique velocity field solution. In Appendix IIb, it is shown that for the simple case of a line segment moving rigidly in space, with known velocity vectors at its endpoints, there does not exist a unique velocity field that minimizes either variation in direction, or variation in magnitude of velocity. This result is not surprising; velocity has two degrees of freedom, and can be decomposed in various ways (for example, into direction and magnitude, $x$ and $y$ components, normal and tangential components, or radial and rotational components). If the measure of variation in velocity constrains only one of these degrees of freedom, such as direction, the other degree of freedom is allowed to vary freely, and cannot be specified uniquely. The simpler functional, $\int|\frac{\partial \mathbf{V}}{\partial s}|ds$, also does not yield a unique velocity field in general. In Appendix IIb, an example is presented for which a velocity field that minimizes this measure does not exist. Measures involving higher order derivatives impose a higher degree of smoothness on the velocity field, which may not be necessary for this computation. Our approach is to first consider the simplest measure, $\int|\frac{\partial \mathbf{V}}{\partial s}|^2 ds$; if this is not sufficient for the velocity field computation, we can then consider a more complex measure.

### 3.3.3. Physical Plausibility of the Velocity Field Solution

The second criterion that can be used to evaluate a particular measure of variation in velocity is the physical plausibility of the resulting solution. One question that can be addressed is, under what conditions will the velocity field that minimizes $\int|\frac{\partial \mathbf{V}}{\partial s}|^2 ds$ be the correct physical velocity field? If orthographic projection of the scene onto the image plane is assumed, there are at least two classes of motion for which this is true. The first consists of rigid objects undergoing pure translation. In this case, $\frac{\partial \mathbf{V}}{\partial s} = 0$ everywhere along contours in the image, and hence $\int|\frac{\partial \mathbf{V}}{\partial s}|^2 ds = 0$. Since zero is the smallest value that the measure can obtain, it follows that if there exists a valid solution (satisfying the constraints derived from the image) that is consistent with pure translation, then this solution minimizes $\int|\frac{\partial \mathbf{V}}{\partial s}|^2 ds$.

The second class of motions includes rigid polyhedra, rotating and translating in space. In Appendix III we show the following:

> *Suppose that a rigid three-dimensional object, consisting of straight lines intersecting in space, projects onto the image plane, using orthographic projection, in such a way that line intersections are preserved (that is, two lines intersect in the image if and only if their generators intersect in space). Further, suppose that this object undergoes a general displacement in space. Then the two-dimensional velocity field that satisfies $v^{\perp}(s)$ measured along contours in the image, and minimizes $\int|\frac{\partial \mathbf{V}}{\partial s}|^2 ds$, is the correct projected two-dimensional velocity field.*

The proof of this result takes advantage of the fact that the normal and tangential components of velocity vary linearly along a straight line that is moving rigidly in space. The class of *blocks world* objects, whose motion was studied by Roach and Aggarwal (1979), is incorporated by the above result. The rigid motion of simple polygons in the image plane (Aggarwal and Duda 1975) is a special case of this class of motions, for which an algorithm that minimizes $\int|\frac{\partial \mathbf{V}}{\partial s}|^2 ds$ is guaranteed to compute the correct velocity field.

More recently, Yuille (1983) derived a general condition under which the velocity field that minimizes $\int|\frac{\partial \mathbf{V}}{\partial s}|^2 ds$ is the correct velocity field. Let $\mathbf{V}'(s)$ denote the true projected two-dimensional velocity field for a contour in motion, and let $\mathbf{T}(s)$ denote the local tangent vector along the contour. If the following relationship holds at every point on the contour:

$$\mathbf{T} \cdot \frac{\partial^2 \mathbf{V}'}{\partial s^2} = 0 \tag{17}$$

then the velocity field $\mathbf{V}(s)$ that satisfies the constraints imposed by $v^{\perp}(s)$ and minimizes $\int |\frac{\partial \mathbf{V}}{\partial s}|^2 ds$ is equal to the true velocity field $\mathbf{V}'(s)$. The two classes of motion mentioned above correspond to cases for which $\frac{\partial^2 \mathbf{V}'}{\partial s^2} = 0$ everywhere along the contour, so that this general condition holds trivially.[1]

For the class of smooth curves, undergoing general motion in space, we rely on an empirical investigation, presented in Chapter 4, for demonstrating the physical plausibility of the resulting solution. For this class of motions, the velocity field of least variation, is in general not the physically correct one. It is often qualitatively similar to the true velocity field, however. When the two velocity fields differ significantly, it appears that the smoothest velocity field may be more consistent with human motion perception; this is examined further in Chapters 4 and 6.

A possible constraint on the velocity field, that is not considered explicitly, is the rigidity of the underlying surface. The computation of a *smoothest* velocity field does not necessarily seek a solution that corresponds to rigid motion, in either two or three dimensions. This may at first seem physically implausible. When a three-dimensional curve rotates in space, however, its two-dimensional projection may undergo significant distortion in the image. As a consequence, it is unreasonable to formulate the measurement of motion in the image as a search for a velocity field that corresponds to rigid motion in two dimensions. Without knowing the three-dimensional structure of the curve, it is very difficult, if not impossible, to find a two-dimensional velocity field that corresponds to a single rigid motion in three dimensions. It is also the case that some of the motion that we encounter arises from surfaces that are nonrigid. If the analysis of motion is a two-stage process, with the measurement of two-dimensional motion preceding the derivation of three-dimensional structure from motion, a constraint such as smoothness may be the most restrictive type of constraint that may be used for the measurement stage, which yields a unique solution, and still allows the analysis of general motion.

---

[1] Yuille (1983) notes that this condition is also satisfied when a curve undergoes pure expansion in the image plane.

### 3.3.4. Alternate Formulations of the Smoothness Constraint

In the theoretical development presented in the previous sections, smoothness of the velocity field was imposed along one-dimensional contours in the image. In this section, three alternate formulations of the smoothness constraint are briefly described. In the first, smoothness is imposed over two dimensions, rather than along contours. The second considers smoothness of the velocity field in three dimensions. Finally, the smoothness constraint is expressed in terms of deformation of contours over time, rather than variation of velocity along contours. The purpose of this section is to convey the variety of ways in which the notion of smoothness may be used in the measurement of motion.

The smoothness constraint can be formulated in two dimensions, rather than along one-dimensional contours. In the work of Horn and Schunck (1981), for example, variation in the velocity field is expressed in terms of the variation in the $x$ and $y$ components of velocity, in two dimensions. In particular, variation is defined by the following measure:

$$\left(\frac{\partial \mathbf{V}_x}{\partial x}\right)^2 + \left(\frac{\partial \mathbf{V}_x}{\partial y}\right)^2 + \left(\frac{\partial \mathbf{V}_y}{\partial x}\right)^2 + \left(\frac{\partial \mathbf{V}_y}{\partial y}\right)^2. \tag{18}$$

Initial motion measurements are obtained wherever the image intensity gradient is nonzero, and a dense velocity field is computed, which simultaneously satisfies the constraints derived from the image, and minimizes the above measure of variation in velocity. An advantage to applying the smoothness constraint over two dimensions is that it yields a dense velocity field, with motion measurements everywhere in the image, rather than only at the location of contours.[2] To apply the smoothness constraint over an area of the image, however, it is necessary to specify a neighborhood size, within which constraints are combined, and smoothness imposed on the velocity field. Unless surface boundaries can be defined prior to the velocity field computation, specifying an appropriate area of the image is difficult. In

[2]When viewing a surface in motion, we certainly have the impression of a complete surface, for which motion is known everywhere. This ultimate dense representation of velocity may be the consequence of an initial dense two-dimensional velocity field computation, as suggested by Horn and Schunck (1981; Schunck 1983). A second possibility is that the measurement of motion, and recovery of three-dimensional structure take place initially at the location of features of the image, and the dense representation is then constructed by a surface interpolation process, that integrates three-dimensional information from multiple visual sources (Grimson 1981; Terzopoulos 1983).

general, the extent of image contours is more highly correlated with single surfaces; if the smoothness constraint is applied to single contours, the problem of integrating motion measurements across object boundaries is reduced.

One could also attempt to impose smoothness of the velocity field in three dimensions, rather than two. If the rate of change of the $z$ component of a curve (the depth component) varies significantly over the curve, then the distribution of variation in the three-dimensional velocity field implied by the smoothest two-dimensional velocity field, may be highly nonuniform. One difficulty with imposing smoothness in three dimensions is that the structure of the curve is not known initially, so that variation cannot be expressed with respect to the three-dimensional curve parameters. Second, the strategy of computing the velocity field of *least* variation may not be appropriate in three dimensions.[3]

The previous formulations of the smoothness constraint are all based on the assumption that a smooth surface in motion gives rise to an instantaneous velocity field that varies smoothly over the surface. There is a different way in which the notion of smoothness may be embodied in the velocity field computation; that is to assume that surfaces deform smoothly over time. With regard to contour-based schemes for motion measurement, rather than deriving a velocity field for which variation along image contours is as small as possible, a velocity field could be derived that represents the least local deformation of the contours over time. One way in which this deformation can be characterized is by the change over time, in the local orientation of corresponding segments of the moving contour. Defined in this way, local deformation could be minimized by requiring that elements along the contour at one moment in time move toward elements at the next moment, that are as similar in orientation as possible (that is, the directions of the local tangents are similar (Yuille 1983)). The idea of minimizing the total change in orientation of elements in motion has been used previously in correspondence schemes for motion measurement (for example, Ullman 1979a).

---

[3]A recent theoretical analysis by Shimon Ullman and Alan Yuille (personal communication), based on previous observations by Ullman (1979a, Section 3.6), suggests that the computation of the smoothest velocity field in two dimensions may be a sound strategy, from a probabilistic viewpoint. The soundness of the strategy is derived from the observation that the projection of three-dimensional objects in motion onto the image plane tends to yield smoother velocity fields in two dimensions.

For the case of continuous motion, the two formulations of the smoothness constraint may yield similar solutions (Yuille 1983). In Section 4.4.1, the velocity field of least variation is illustrated, for a number of smooth curves in motion. The solutions appear to have the property that points on the curve at a particular orientation (that is, with a particular tangent direction) move toward points on the curve at the next moment in time, which are of the same orientation. When a curve rotates, the real corresponding elements differ in orientation; if the velocity field computation attempts to link elements of similar orientation over time, the resulting solution may imply a global distortion of the curve. Formulation of the smoothness constraint in terms of variation in velocity along contours was useful for demonstrating theoretical properties of the velocity field computation, such as the uniqueness of the solution, and correctness of the solution for particular classes of motion. A formulation in terms of the local deformation of contours over time may be useful for the design of a specific algorithm for computing the velocity field.

### 3.4. Deriving Additional Constraint from the Changing Image

In the previous section, two sources of constraint were used for the velocity field computation. Constraint from the image consisted of a single curve at a particular moment in time, together with the instantaneous measurements of the perpendicular component of velocity along the curve. As an additional source of constraint, we considered the computation of the velocity field that exhibits the least variation along the curve. Further constraint can be derived from the changing image if the computation is not restricted to the use of instantaneous measurements. For example, a second curve, at some time later, could be used as a source of constraint. If the time interval is small, then the displacement of points between the two curves is also small. It can then be required that each point on the first curve project to a point on the second, with a velocity that is consistent with the instantaneous perpendicular component of velocity $v^{\perp}(s)$. The use of the second curve in this way relies on the assumption that velocity is constant over the time interval separating the two curves. In addition, the availability of a second curve also allows for the tracking of identifiable features, such as points of high curvature (Lawton 1983).

This approach may yield a simpler, more robust algorithm for the velocity field computation, because it utilizes more constraint from the image. It has the disadvantage, however, that theoretical results regarding the uniqueness of the

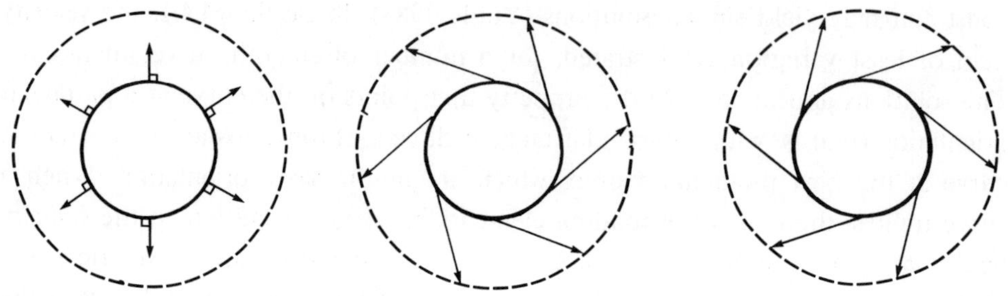


**Figure 3.8. Ambiguity of the Velocity Field.** (a) Initial constraints on the velocity field; the arrows indicate the perpendicular components of velocity, and the dotted line indicates the second curve. (b) and (c) Two possible velocity field solutions corresponding to expansion and rotation of the circle to the left and right, respectively.

solution may be more difficult to obtain. A simple example, in which the velocity field solution is ambiguous, is shown in Figure 3.8. Suppose that the initial constraints shown in Figure 3.8a are given. The arrows indicate the perpendicular components of velocity along the first curve, and the dotted line indicates the second curve. If the velocity field is forced to satisfy both sources of constraint, there are two possible velocity field solutions, shown in Figures 3.8b and c. The solutions correspond to the two directions of rotation of the circle. The local velocity vectors intersect the second curve, and have a component in the direction perpendicular to the first curve that agrees with the perpendicular components given in Figure 3.8a. Both velocity fields also exhibit the same total variation. The use of instantaneous motion measurements alone, together with the additional smoothness constraint, as discussed in Section 3.3, yields the single velocity field given by the vectors in Figure 3.8a, corresponding to pure expansion of the circle. The additional constraint, imposed by forcing the velocity field to intersect the second curve, leads to an ambiguous situation. In general, theoretical results on uniqueness may be more difficult to obtain for this integrated approach to the velocity field computation.

The availability of a second curve, at some time later, may simplify the velocity field computation in the following way. The perpendicular component of velocity, measured at a point $p$ on the first curve, constrains the velocity vector at $p$ to project to a point along the line $l$ in the next time frame, shown in Figure 3.9. If in addition, $p$ must project to a point $q$ on the second curve, possible candidates for $q$ may be

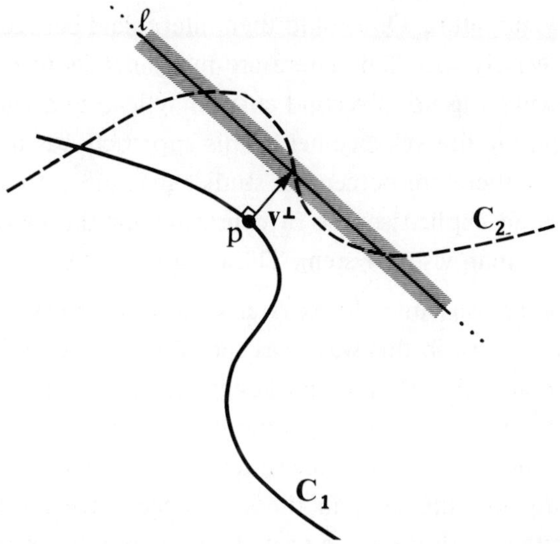


**Figure 3.9. Using a Second Curve.** The perpendicular component of velocity at the point $p$ on $\mathbf{C}_1$ restricts the velocity at $p$ to project to the line $l$. Error in this measurement extends $l$ to the grey band shown. Adding the constraint of the second curve $\mathbf{C}_2$, possible velocity vectors must now lie within the intersection of the grey band with $\mathbf{C}_2$.

given by the intersection of $l$ with the second curve. In practice, $v^{\perp}(s)$ may not be constant over small time intervals, and there will be error in its measurement. As a consequence, a band in the second frame, to which $p$ must project, should be considered, as shown in Figure 3.9. Candidates for $q$ are then given by the intersection of this band with the second curve. If the curve has fairly high local curvature, or undergoes rotation, then this intersection alone provides considerable constraint on the possible local velocity vectors. In the worst case of an extended line undergoing pure translation, the second curve offers limited additional constraint. We noted in Section 3.3.3 that velocity varies linearly along a straight line. The combination of this observation, and the use of the second curve described here, suggests a fairly simple velocity field algorithm. First, the segments of the second curve to which each point on the first curve must project can be computed, using the intersection scheme illustrated in Figure 3.9. (In addition, elements of the two curves that are similar in orientation, curvature, or other properties could be matched, further restricting the local velocity vectors.) After this stage, velocity would be highly constrained around areas of high curvature, and only weakly constrained around areas of low curvature. One could then interpolate between points at which velocity is more precisely specified. There are many details to be examined more carefully, but the explicit use of a second curve may lead to a fairly simple, robust algorithm for computing the velocity field. This approach was not pursued in this work, in part because there are perceptual studies that suggest that a second curve may not be used as an explicit source of constraint for the motion measurement computation in the human visual system. These studies are examined in Chapter 6.

The assumption of constant velocity over small time intervals, which underlies the use of the second curve in this way, was shown to be useful for spatiotemporal interpolation in human vision (Fahle and Poggio 1981). We mentioned in Chapter 2 that the human system is capable of "filling in" or interpolating, between stimuli undergoing discrete motion. The spatial and temporal properties of the initial filters in human vision are not sufficient, in theory, to reconstruct visual patterns that are presented discretely, with the large spatial and temporal displacements that are demonstrated psychophysically. Fahle and Poggio (1981) show, however, that with the additional assumption of constant velocity over time, properties of the initial filters are sufficient to reconstruct the visual input

### 3.5. Summary

Various additional constraints have been considered for the computation of the velocity field from initial motion measurements derived from the changing image. These constraints range from the restricted assumption of pure translation to the general constraint of smoothness of the velocity field, which allows the analysis of the projected motion of arbitrary rigid and nonrigid surfaces in space. The use of these different constraints results in considerable variation in the classes of motion that may be analyzed, the type of algorithm that may be used, and the extent of the theoretical analysis required to formulate a well-defined computational problem. In analyzing these constraints, we have so far addressed purely computational issues. In the next chapter, we continue this focus, discussing possible algorithms for computing the velocity field, which utilize the smoothness constraint.

# 4. AN ALGORITHM TO COMPUTE THE VELOCITY FIELD

In the previous chapter, the use of the smoothness constraint led to the formulation of the velocity field computation as an optimization problem. We seek a velocity field solution that satisfies the constraints derived from the changing image, and minimizes the measure of variation along contours given by: $\int|\frac{\partial \mathbf{V}}{\partial s}|^2 ds$. The computation may also be described as seeking a solution for which neighboring velocity vectors are as similar as possible. To further test the adequacy of this approach, we can specify an algorithm that embodies the smoothness constraint, and examine the results of the algorithm for a number of motion sequences. An algorithmic study first allows us to demonstrate the biological feasibility of this approach, by showing that this formulation of the velocity field computation lends itself naturally to algorithms that involve simple, local, parallel operations (Ullman 1979b; Grimson 1981; Marr 1982). Second, an empirical study provides a means for testing the physical plausibility of the resulting velocity field solution. While the velocity field across a physical surface is generally expected to be smooth, it is not yet clear whether the velocity field of *least* variation is still physically plausible.

In general, there are many algorithms that implement a given computational theory. For the velocity field computation, there are many choices for the design of an algorithm that embodies the additional smoothness constraint. In this chapter and in Appendix IV, several standard algorithms are presented that may be used to solve the optimization problem posed. To implement the computation, we chose an iterative algorithm from mathematical programming, known as the *conjugate gradient algorithm* (Luenberger 1973). This particular algorithm is certainly not appropriate as a model for human vision. Our aim is to test the basic idea of computing the velocity field of least variation. If the results of the algorithm support the feasibility of this idea, we can then explore alternative algorithms to implement the theory, that may be more appropriate for the human system. This separability of the computational theory and algorithmic levels of analysis is an important aspect of the computational approach to the study of visual processing.

## 4.1. The Discrete Formulation

In Chapter 3, a continuous formulation of the smoothness constraint was presented; the image, however, is discrete, and therefore image contours consist of a set of discrete points. The first step in the design of an algorithm is to convert the continuous formulation into a discrete one. To do this, the functional is first expressed in terms of the $x$ and $y$ components of velocity, which we denote by $\mathbf{V}_x$ and $\mathbf{V}_y$.[1] The continuous functional becomes:

$$\Theta = \int \left[ \left( \frac{\partial \mathbf{V}_x}{\partial s} \right)^2 + \left( \frac{\partial \mathbf{V}_y}{\partial s} \right)^2 \right] ds. \tag{18}$$

Suppose there are $n$ points along a contour, whose coordinates are given by:

$$\{(x_1, y_1), (x_2, y_2), ..., (x_n, y_n)\}.$$

Let $(\mathbf{V}_{x_i}, \mathbf{V}_{y_i})$ denote the velocity at the point $(x_i, y_i)$. Assuming for now that the contour is closed, and the points are evenly spaced along the contour, the following function defines a discrete correlate to the continuous functional:

$$\Phi_1 = \sum_{i=2}^{n} \left[ (\mathbf{V}_{x_i} - \mathbf{V}_{x_{i-1}})^2 + (\mathbf{V}_{y_i} - \mathbf{V}_{y_{i-1}})^2 \right] + \left[ (\mathbf{V}_{x_1} - \mathbf{V}_{x_n})^2 + (\mathbf{V}_{y_1} - \mathbf{V}_{y_n})^2 \right]. \tag{19}$$

For the case of an open contour, the discrete differences at the endpoints change. If points on the contour are not evenly spaced, the discrete differences must be divided by the distances between adjacent points on the contour. The goal of the computation is to compute the set of $x$ and $y$ components of velocity, $\{(\mathbf{V}_{x_1}, \mathbf{V}_{y_1}), (\mathbf{V}_{x_2}, \mathbf{V}_{y_2}), ..., (\mathbf{V}_{x_n}, \mathbf{V}_{y_n})\}$, which minimizes the discrete function $\Phi_1$.

### 4.2. Satisfying the Image Constraints Exactly

The second step toward defining an algorithm is to incorporate the constraints on velocity that are derived from the changing image. Here there is a choice; the velocity field can be forced to satisfy the constraints exactly, or satisfy them approximately. In the first case, the constraints can be expressed as follows:

$$\mathbf{V} \cdot \mathbf{u}^{\perp} - v^{\perp} = 0. \tag{20}$$

[1] Initial investigation indicated that this decomposition of velocity leads to simpler algorithms, than a decomposition into perpendicular and tangential components: details are given in Appendix IVa.

The equation states explicitly that the perpendicular component of velocity for the computed velocity field, $\mathbf{V} \cdot \mathbf{u}^{\perp}$, is equal to the measured perpendicular component, $v^{\perp}$. (For simplicity, we omit the argument, $s$, to these functions.) Letting $\mathbf{V} = (\mathbf{V}_x, \mathbf{V}_y)$ and $\mathbf{u}^{\perp} = (\mathbf{u}_x^{\perp}, \mathbf{u}_y^{\perp})$, the constraints are simple linear constraints:

$$\mathbf{V}_{x_i}\mathbf{u}_{x_i}^{\perp} + \mathbf{V}_{y_i}\mathbf{u}_{y_i}^{\perp} - v_i^{\perp} = 0. \tag{21}$$

This equation can be used to express the $y$ components of velocity in terms of the $x$ components:

$$\mathbf{V}_{y_i} = \frac{v_i^{\perp} - \mathbf{V}_{x_i}\mathbf{u}_{x_i}^{\perp}}{\mathbf{u}_{y_i}^{\perp}}. \tag{22}$$

This expression is valid for $\mathbf{u}_{y_i}^{\perp} \neq 0$. If $\mathbf{u}_{y_i}^{\perp} = 0$, then $\mathbf{u}_{x_i}^{\perp} = 1$, and it follows that $\mathbf{V}_{x_i} = v_i^{\perp}$, and $\mathbf{V}_{y_i}$ is unknown. This situation arises at points on the contour for which the vector $\mathbf{u}^{\perp}$ is oriented horizontally (the contour is oriented vertically). At such a point, $v^{\perp}$ specifies the $x$ component of velocity directly, and the $y$ component is unconstrained. For points at which $\mathbf{u}_{y_i}^{\perp} \neq 0$, the above expressions for $\mathbf{V}_{y_i}$ can be substituted directly into the discrete function $\Phi_1$. At points for which $\mathbf{u}_{y_i}^{\perp} = 0$, the known values of $\mathbf{V}_{x_i}$ and unknown parameters $\mathbf{V}_{y_i}$ can be substituted into $\Phi_1$. Given $n$ points on the contour, there remains $n$ unknown parameters to compute, with no further constraint on these parameters. The goal of the computation in this case is to find the set of $n$ parameters that minimizes the new discrete objective function, which we denote by $\Phi_2$. This set of parameters satisfies the system of $n$ linear equations given by:

$$\frac{\partial \Phi_2}{\partial \mathbf{V}_{x_i}} = 0, \qquad 1 \leq i \leq n, \quad \mathbf{u}_{y_i}^{\perp} \neq 0; \tag{23}$$

$$\frac{\partial \Phi_2}{\partial \mathbf{V}_{y_i}} = 0, \qquad 1 \leq i \leq n, \quad \mathbf{u}_{y_i}^{\perp} = 0. \tag{24}$$

We show in Appendix IVa that this system of equations has a unique solution, except for the case in which the image consists of an extended straight line.

There are many possible algorithms for solving such a system of linear equations. Two examples are the Jacobi and Gauss-Seidel relaxation methods. Both are iterative methods that compute a sequence of approximations to the final solution, given an initial approximation, in which the parameters are usually assigned the value

0. These methods both involve simple, local operations. The Jacobi method is inherently parallel, but exhibits slower convergence than the Gauss-Seidel method, which is serial. Details of the application of the Jacobi and Gauss-Seidel relaxation methods to the velocity field computation, for the case in which image constraints are to be satisfied exactly, are given in Appendix IVa.

### 4.3. Satisfying the Image Constraints Approximately

In general, there will be error in the measurements of $v^{\perp}$. From a practical standpoint, it may be advantageous to require that the velocity field only approximately satisfy these image constraints. This can be accomplished by requiring that the difference between $\mathbf{V} \cdot \mathbf{u}^{\perp}$ and the measured $v^{\perp}$ be small. The continuous functional can be extended as follows:

$$\Theta = \int \left[ \left( \frac{\partial \mathbf{V}_x}{\partial s} \right)^2 + \left( \frac{\partial \mathbf{V}_y}{\partial s} \right)^2 \right] ds + \beta \int \left[ \mathbf{V} \cdot \mathbf{u}^{\perp} - v^{\perp} \right]^2 ds. \qquad (25)$$

$\beta$ is a weighting factor that expresses our confidence in the measured velocity constraints. (In this particular formulation, $\beta$ is constant over the entire contour, but in principle, $\beta$ could vary, if our confidence in the precision of individual measurements of $v^{\perp}$ varies along the contour.) The second term describes the squared difference between the computed and measured perpendicular components of velocity. There will also be error in the measurement of the direction vector $\mathbf{u}^{\perp}$; some of this error is captured implicitly by the second term of the above functional. The continuous functional leads to the following discrete function:

$$\Phi_3 = \Phi_1 + \beta \sum_{i=1}^{n} \left[ \mathbf{V}_{x_i} \mathbf{u}_{x_i}^{\perp} + \mathbf{V}_{y_i} \mathbf{u}_{y_i}^{\perp} - v_i^{\perp} \right]^2. \qquad (26)$$

The goal of the computation in this case is to find the set of $x$ and $y$ components of velocity, $\{(\mathbf{V}_{x_1}, \mathbf{V}_{y_1}), (\mathbf{V}_{x_2}, \mathbf{V}_{y_2}), \ldots, (\mathbf{V}_{x_n}, \mathbf{V}_{y_n})\}$, which minimizes the discrete function $\Phi_3$. These components satisfy the system of $2n$ linear equations given by:

$$\frac{\partial \Phi_3}{\partial \mathbf{V}_{x_i}} = 0, \qquad \frac{\partial \Phi_3}{\partial \mathbf{V}_{y_i}} = 0, \qquad 1 \leq i \leq n. \qquad (27)$$

We could construct this set of linear equations, and solve them using methods such as Jacobi and Gauss-Seidel relaxation, as in the previous section. In general, however, there may be several hundred points on a given contour, giving rise to a system of several hundred linear equations. The Gauss-Seidel and Jacobi methods exhibit slow convergence for such a large system of equations, and only converge asymptotically to the final solution. As a consequence, we instead turn to an algorithm from mathematical programming for solving this optimization problem.

The general mathematical programming problem can be stated as:

$$\begin{array}{lll} \text{minimize} & f(\mathbf{x}) & \\ \text{subject to} & h_i(\mathbf{x}) = 0 & i = 1, ..., m \\ & g_j(\mathbf{x}) \leq 0 & j = 1, ..., r \\ & \mathbf{x} \in S & \end{array}$$

where $\mathbf{x}$ is an n-dimensional vector of unknowns, $\mathbf{x} = (x_1, x_2, ..., x_n)$. The objective function $f$, and constraints $h_i$, $i = 1, ..., m$ and $g_j$, $j = 1, ..., r$ are real-valued functions of the variables $x_1, x_2, ..., x_n$. The set $S$ is a subset of the n-dimensional space. A particular optimization problem may or may not give rise to constraints of the form $h_i(\mathbf{x}) = 0$ or $g_j(\mathbf{x}) \leq 0$; that is, the problem may be a *constrained* or *unconstrained* optimization problem.

For the case in which the image constraints are satisfied approximately, the velocity field computation can be formulated as an unconstrained optimization problem. In this case, we let:

$$\mathbf{x} = \{\mathbf{V}_{x_1}, \mathbf{V}_{x_2}, ..., \mathbf{V}_{x_n}, \mathbf{V}_{y_1}, \mathbf{V}_{y_2}, ..., \mathbf{V}_{y_n}\}.$$

The objective function $f(\mathbf{x}) = \Phi_3$. We seek a vector $\mathbf{x}$ that minimizes $f(\mathbf{x})$, with no further constraints on $\mathbf{x}$.

For the particular examples presented in this chapter, the only constraints obtained from the image are the measurements of $v^{\perp}$ along contours. In general, there may be points at which $\mathbf{V}(s)$, or the direction of velocity alone, is known. This information may be the result of a process that tracks localizable features in the image, such as corners, isolated points, or line endpoints. These additional constraints may be incorporated easily into the velocity field algorithm. As before, the constraints may be satisfied exactly or approximately. Let $(\mathbf{C}_{x_i}, \mathbf{C}_{y_i})$ denote the measured two-dimensional velocity vector at a particular point $(x_i, y_i)$. For the case in which the constraints are satisfied exactly, the values $\mathbf{C}_{x_i}$ and $\mathbf{C}_{y_i}$ can be substituted directly for $\mathbf{V}_{x_i}$ and $\mathbf{V}_{y_i}$ in the set of linear equations derived from $\Phi_3$, thereby reducing the total number of equations by two. If only the direction of velocity is known at a point, the $x$ and $y$ components of velocity are related by an equation of the form $\mathbf{V}_{y_i} = c_i \mathbf{V}_{x_i}$ for some constant $c_i$. By substituting this expression for $\mathbf{V}_{y_i}$ into $\Phi_3$, the total number of resulting equations may be reduced by one. If these additional constraints are to be satisfied approximately, then the difference between the known and computed parameters can be incorporated into the second term of the discrete function $\Phi_3$. Thus, further velocity information may be incorporated easily into the velocity field algorithm.

There are several standard algorithms from mathematical programming that can be used to solve this particular optimization problem (see, for example, Luenberger 1973). For the examples in the next section, the conjugate gradient algorithm was used to obtain the solution. This is an iterative algorithm that utilizes the gradient of the objective function $f(\mathbf{x})$ to choose an optimal path to follow toward the final solution. The initial velocity field is given by the perpendicular velocity vectors, $v^{\perp}\mathbf{u}^{\perp}$, along a contour. If there are $n$ parameters to be computed, the conjugate gradient algorithm is guaranteed to converge to the exact solution in at most $n$ iterations. Details of the application of the conjugate gradient algorithm to the velocity field computation are given in Appendix IVb.

### 4.4. Examples of the Smoothest Velocity Field

In this section, we present examples of the velocity field of least variation, for a variety of motion sequences. The conjugate gradient algorithm, described in Appendix IVb, was chosen for its fast convergence and because it is guaranteed to compute the velocity field of least variation. The algorithm provides us with a means of testing whether the use of the smoothness constraint is a reasonable approach to the velocity field computation.

#### 4.4.1. Ideal Smooth Curves

We begin with some simple curves, undergoing rigid motion. For this first set of examples, the curves and the initial perpendicular components of velocity were generated analytically from a known velocity field, and therefore represent ideal input for the algorithm. Many of the examples were chosen because perceptual studies indicate that human observers see motions that differ from the true motion

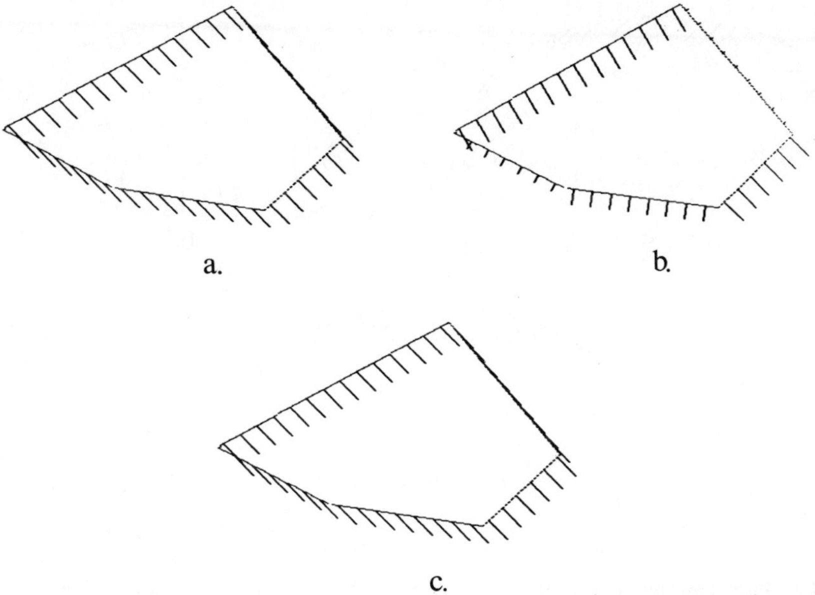


**Figure 4.1. Pure Translation.** (a) The arrows represent a sampling of the true velocity field for the polygon translating in the image plane. (b) The initial perpendicular components of velocity. (c) The computed velocity field of least variation.

of the curves. In Chapter 6, the results of the algorithm presented here are discussed from a perceptual point of view.

### 1. Rigid translation in the image plane

Figures 4.1a and 4.2a show a sampling of the true velocity field for two curves, a polygon and an ellipse, translating rigidly in the image plane. Figures 4.1b and 4.2b illustrate the perpendicular velocity vectors $v^{\perp}\mathbf{u}^{\perp}$, which form the input to the algorithm. As expected, the velocity fields of least variation, displayed in Figures 4.1c and 4.2c, are the correct velocity fields.

### 2. Polygon rotating in the image plane

Figure 4.3 illustrates the true velocity field, initial velocity vectors, and smoothest velocity field for a polygon rotating rigidly in the image plane, about the point O. From the results of Section 3.3.3, we expect to obtain the correct velocity field in this case. Figure 4.3c shows that the smoothest velocity field, as predicted by the

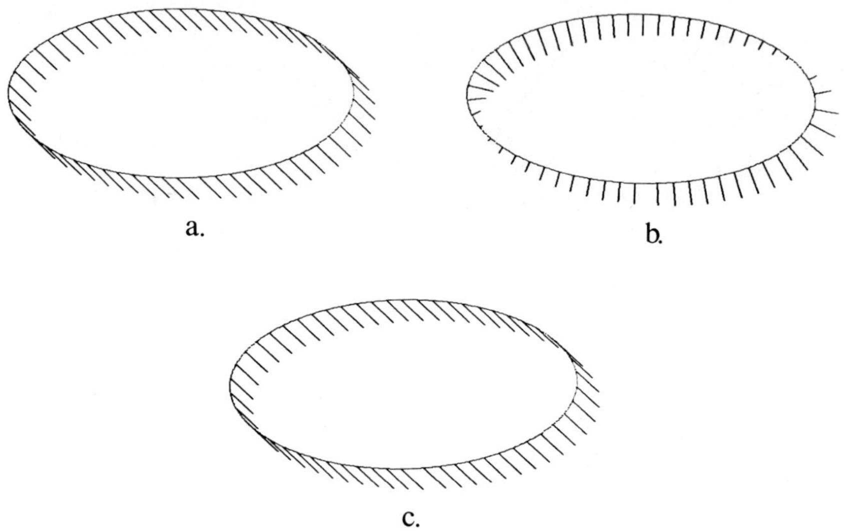


**Figure 4.2. Pure Translation.** (a) The true velocity field for an ellipse translating in the image. (b) The initial perpendicular components of velocity. (c) The computed velocity field of least variation.

conjugate gradient algorithm, is in fact the correct one. This solution depends on the choice of the weighting factor $\beta$ in the functional $\Phi_3$. In this case, $\beta = 0.1$. If $\beta$ is an order of magnitude smaller, the solution differs quantitatively from the true velocity field. In particular, the rotational component of velocity decreases.

Local motion schemes that rely entirely on the intersection of velocity constraints in velocity space (Fennema and Thompson 1979; Marr and Ullman 1981; Thompson and Barnard 1981; Adelson and Movshon 1982; Schunck 1983) cannot derive the velocity field along the extended line segments of the rotating polygon, because the constraints imposed by $v^{\perp}$ form a set of parallel lines in velocity space. Only velocity vectors at the vertices can be obtained directly; additional processing is required to derive the velocity field along the line segments. Velocity varies linearly along straight lines; thus, this additional processing could simply be a linear interpolation of velocity information between the vertices.

### 3. Rotating ellipse

Figure 4.4 illustrates the true velocity field, initial velocity vectors, and smoothest velocity field for an ellipse rotating rigidly in the image plane, about the point O.

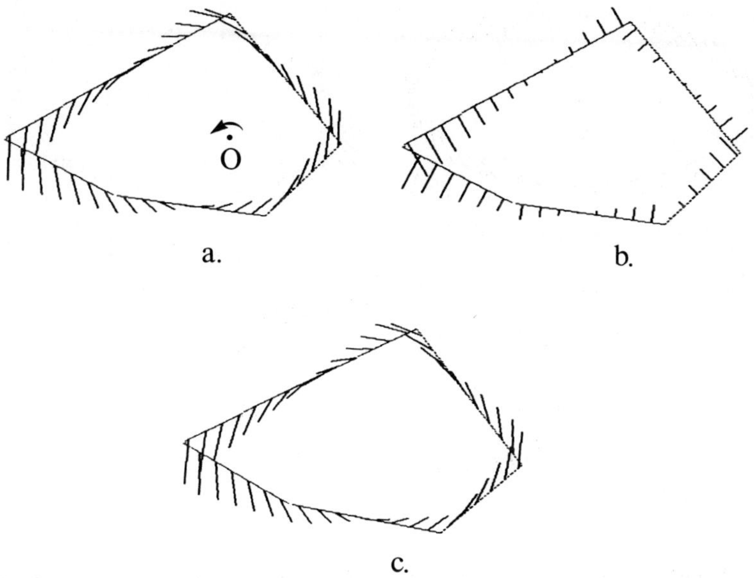


**Figure 4.3. Rotating Polygon.** (a) The true velocity field for a polygon rotating in the image about the point O. (b) The initial perpendicular components of velocity. (c) The computed velocity field of least variation.

In Figure 4.4c, it can be seen that the predicted velocity field is quite different from the true velocity field shown in Figure 4.4a. There is a reduced rotational component of velocity, and added radial component in the computed velocity field. The difference in total variation is significant in this case. For the true velocity field, the discrete measure $\Phi_3 = 0.24$ over the curve, while for the smoothest velocity field, $\Phi_3 = 0.079$. The qualitative difference between the true and smoothest velocity field does not depend critically on the value of $\beta$, over a range from 0.001 to 0.1.

While the smoothest velocity field that is derived in this case is not the correct one, it may be more consistent with human motion perception. An ellipse such as this does not appear rigid when it rotates (Wallach, Weisz and Adams 1956); it appears to deform continuously. In Chapter 6, we show that the perceived movement of a rotating ellipse is qualitatively consistent with the velocity field of least variation, over a wide range of aspect ratios for the ellipse.

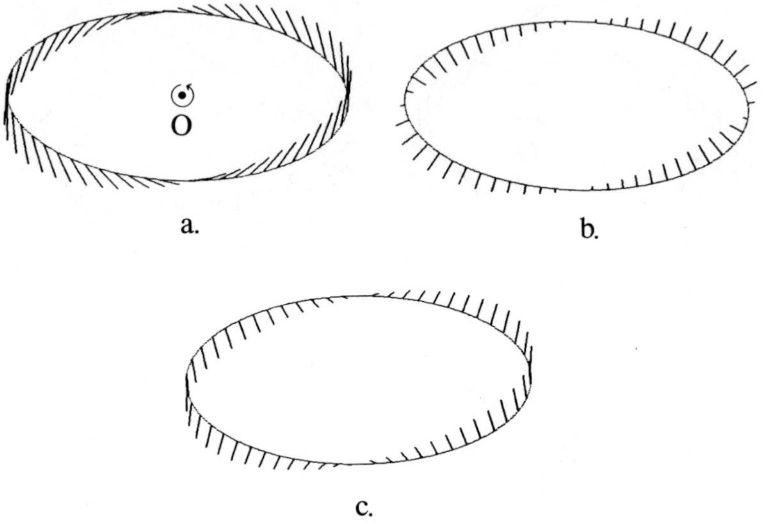


Figure 4.4. **Rotating Ellipse.** (a) The true velocity field for an ellipse rotating in the image about the point O. (b) The initial perpendicular components of velocity. (c) The computed velocity field of least variation.

## 4. Rotating spiral

Figure 4.5 illustrates the true velocity field, initial velocity vectors, and smoothest velocity field for a segment of a logarithmic spiral, rotating rigidly in the image plane, about the point O. While the true motion is pure rotation, the smoothest velocity field exhibits a large radial component of velocity. This radial component decreases with increasing distance from the center of the spiral. The smoothest velocity field in this case is again more consistent with the perception of movement of a rotating spiral, than the true velocity field; a spiral appears to expand or contract when it rotates (see, for example, Holland 1965).

## 5. A deformed circle in rotation

Figure 4.6a shows a deformed circle, undergoing pure rotation in the image plane, about the point O. The true velocity field and perpendicular velocity vectors are shown in Figures 4.6b and c. The perpendicular component of velocity is zero around the circle, due to its symmetry. The computed smoothest velocity field is shown in Figure 4.6d. To better illustrate the true and computed velocity fields, they

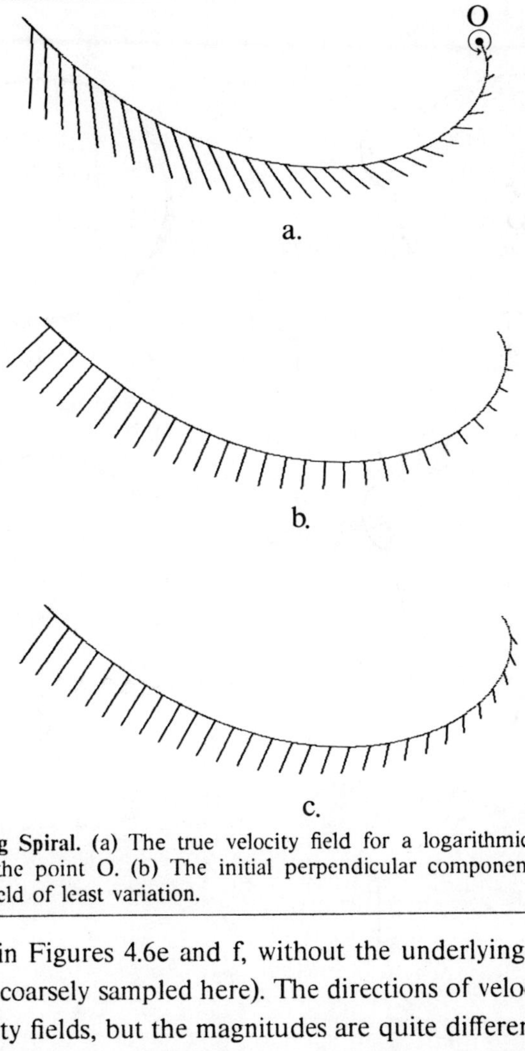


**Figure 4.5. Rotating Spiral.** (a) The true velocity field for a logarithmic spiral rotating in the image plane about the point O. (b) The initial perpendicular components of velocity. (c) The computed velocity field of least variation.

are shown again in Figures 4.6e and f, without the underlying curve (the velocity field is also more coarsely sampled here). The directions of velocity are very similar for the two velocity fields, but the magnitudes are quite different. In the smoothest solution, the magnitude of velocity decreases significantly as distance from the area of the bump increases. Perceptual studies also suggest a consistency with human motion perception for this example (Wallach, Weisz and Adams 1956).

Without a feature on the circle, such as the bump, there is nothing to indicate that the circle is rotating, so the smoothest velocity field would be the zero field (ie, $(V_x, V_y) = (0,0)$ around the circle). The movement of a feature on the circle

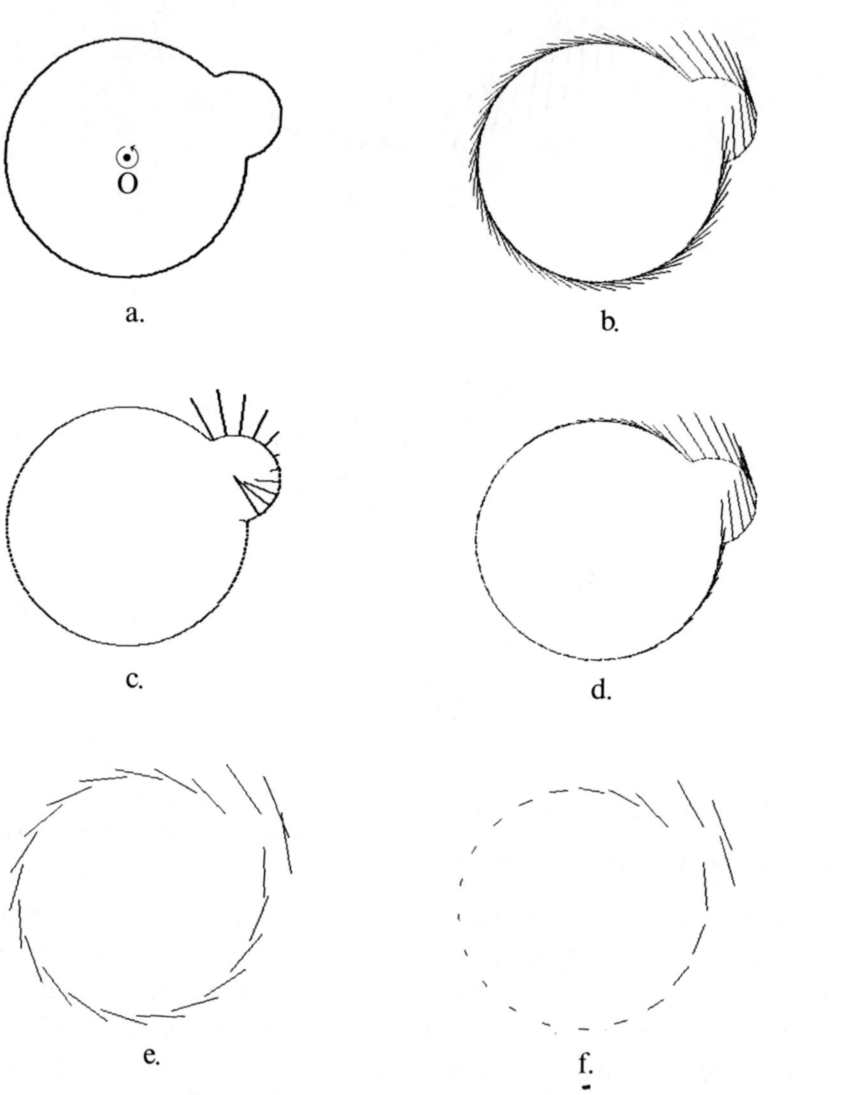


**Figure 4.6. The Deformed Circle.** (a) A deformed circle undergoes pure rotation about the point O. (b) The true velocity field. (c) The perpendicular velocity vectors. (d) The computed velocity field of least variation. (e) A coarse sampling of the true velocity field, without the underlying curve. (f) A coarse sampling of the computed velocity field of least variation.

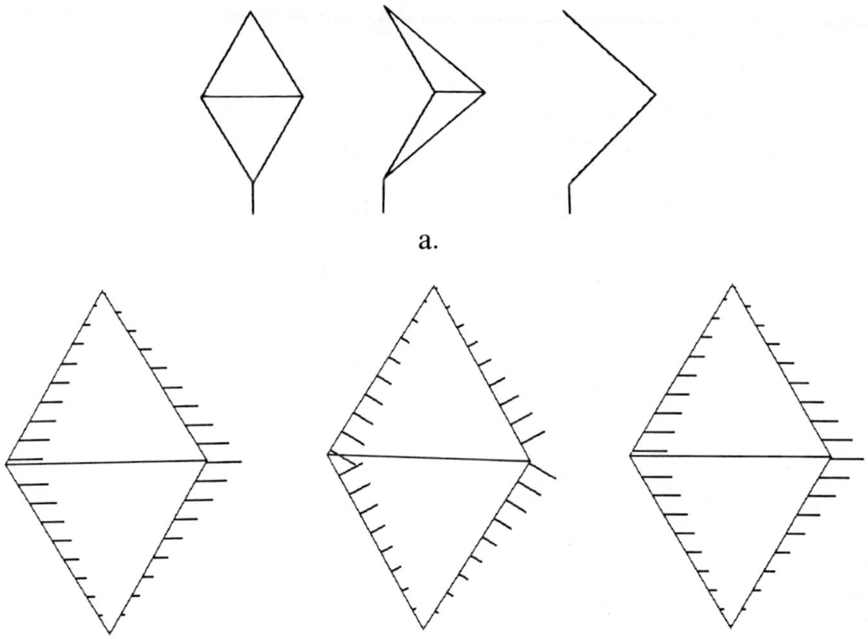

a. b. c. d.

**Figure 4.7. The Kinetic Depth Effect.** (a) A three-dimensional object used by Wallach and O'Connell (1953), shown from three different viewpoints. (b) The true projected velocity field that results from rotation of the figure about the vertical axis. (c) The initial perpendicular components of velocity. (d) The computed velocity field of least variation.

constrains the movement of the rest of the circle, but the propagation of this constraint is spatially limited.

## 6. The kinetic depth effect

Figure 4.7a illustrates a three-dimensional stimulus that was first used by Wallach and O'Connell (1953) to demonstrate the ability of the human visual system to derive three-dimensional structure from the projected two-dimensional motion of an object. They referred to this phenomenon as the *Kinetic Depth Effect.* In their experiments, a three-dimensional wire figure was rotated about its vertical axis, and projected onto a translucent screen. From the two-dimensional projection alone, viewed from the other side of the screen, observers could correctly identify the three-dimensional structure of the figure. Three views of one stimulus are shown

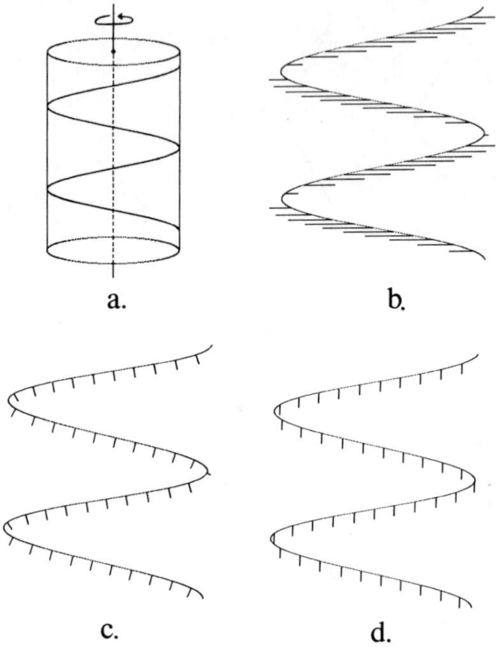


**Figure 4.8. The Barberpole Illusion.** (a) A circular helix on an imaginary cylinder, rotating about the central vertical axis of the cylinder. (b) The two-dimensional projection of the helix and its true velocity field. (c) The initial perpendicular components of velocity. (d) The computed velocity field of least variation.

in Figure 4.7a. The projected front view (using orthographic projection), and true projected velocity field that results when this figure is rotated, are displayed in Figure 4.7b. The initial perpendicular velocity vectors, and computed smoothest velocity field (in two dimensions) are shown in Figures 4.7c and d, respectively. As expected from the theoretical result of Section 3.3.3, the solution in this case is the correct one. The perception of the correct three-dimensional structure of this figure suggests that the human visual system also derives its correct two-dimensional motion initially.

### 7. The barberpole illusion

Figure 4.8a shows a circular helix on an imaginary cylinder, rotating about its central vertical axis. The projection of the curve onto the image plane, together with the resulting two-dimensional velocity vectors, are shown in Figure 4.8b.

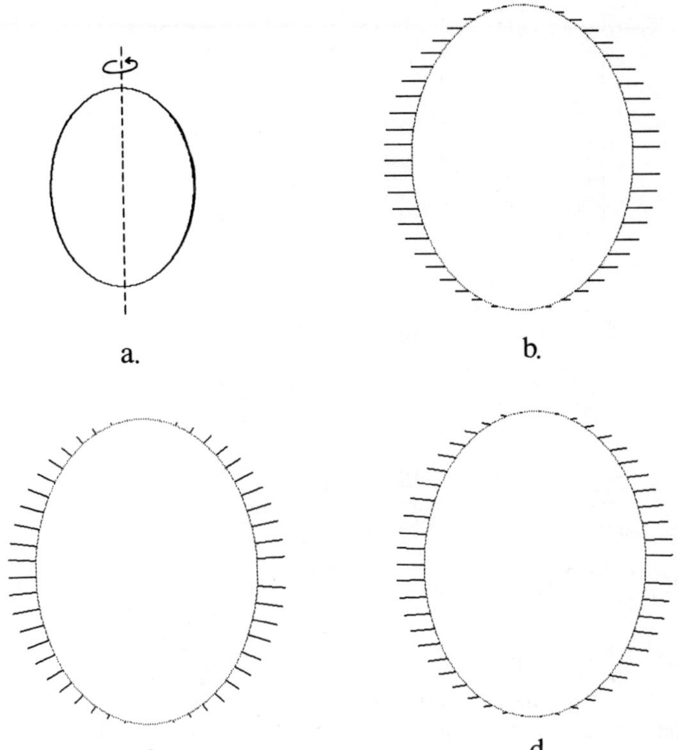


**Figure 4.9. Circle Rotating in Depth.** (a) A circle, slanted 45° away from the image plane, rotating around its central vertical axis. (b) The two-dimensional projection of the circle and its true velocity field. (c) The initial perpendicular velocity vectors. (d) The computed velocity field of least variation.

Figures 4.8c and d show the initial perpendicular velocity vectors, and computed two-dimensional velocity field. The true velocity vectors are strictly horizontal, while those for the smoothest velocity field, which actually correspond to pure translation of the curve, are strictly vertical. The motion of this curve is analogous to the moving stripes of a barberpole, which appear to move downward as the pole rotates.

**8. Circle rotating in depth**

Figure 4.9a shows a circle slanted 45° in depth from the image plane, rotating about the central vertical axis. The projected two-dimensional curve and its velocity vectors are shown in Figure 4.9b. The perpendicular velocity vectors and smoothest

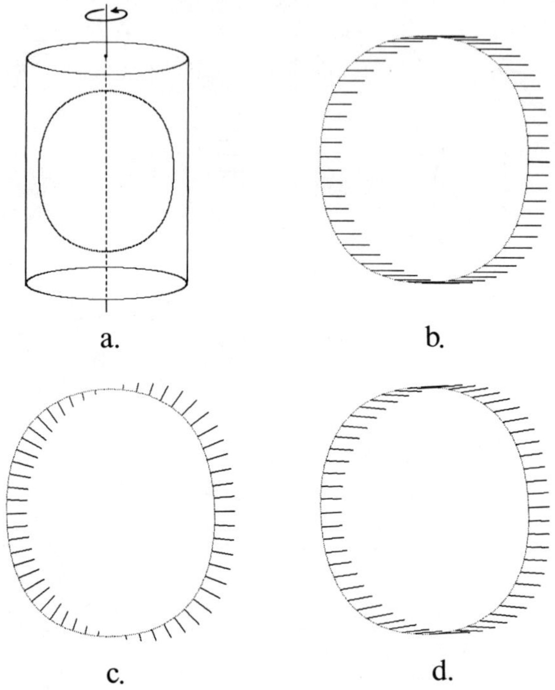


**Figure 4.10. Smooth Curve Rotating in Depth.** (a) A circle, wrapped around an imaginary cylinder, rotating around the central vertical axis of the cylinder. (b) The two-dimensional projection of the curve and its true velocity field. (c) The initial perpendicular components of velocity. (d) The computed velocity field of least variation.

velocity field are shown in Figures 4.9c and d. For the smoothest velocity field, the local velocity vectors deviate by as much as 10° from the strictly horizontal direction of the true velocity vectors. The total variations for the true and smoothest velocity fields are very small ($\Phi_3 = 0.057$ for the true velocity field, and $\Phi_3 = 0.054$ for the smoothest velocity field), and differ by only 5%.

## 9. Smooth curve rotating in depth

In Figure 4.10a, a circle is wrapped around an imaginary cylinder, and rotated about the central vertical axis of the cylinder. The projected two-dimensional curve and its velocity vectors are shown in Figure 4.10b. The perpendicular velocity vectors and smoothest velocity field are shown in Figures 4.10c and d. Figures 4.11 and 4.12 illustrate the velocity fields for this same curve, at different positions in its

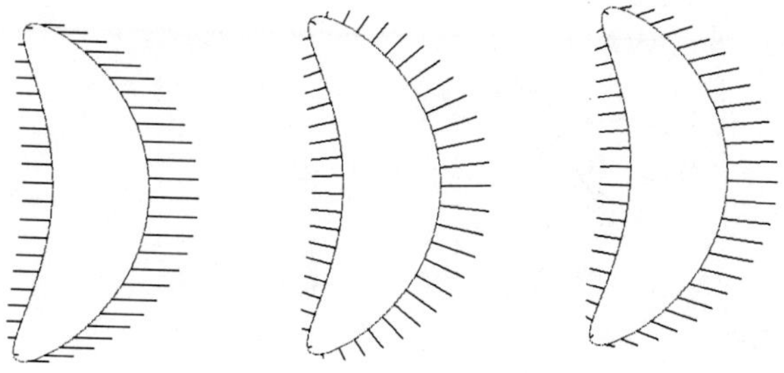

a. b. c.

**Figure 4.11. Smooth Curve Rotating in Depth.** The curve of Figure 4.10 is rotated 70° from its initial position, and then undergoes an instantaneous rotation about the central vertical axis of the cylinder. (a) The two-dimensional projection of the curve and its true velocity field. (b) The initial perpendicular components of velocity. (c) The computed velocity field of least variation.

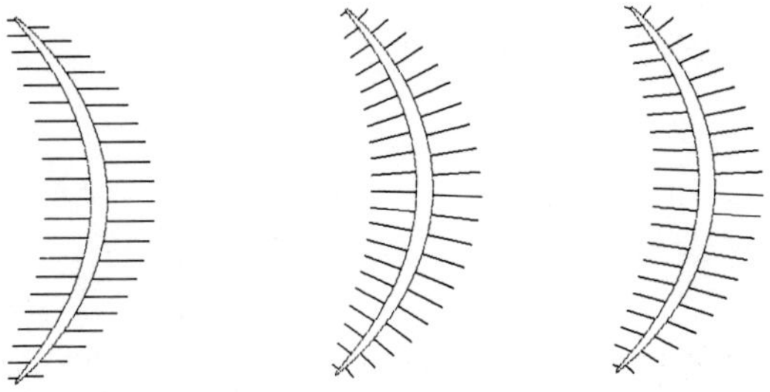

a. b. c.

**Figure 4.12. Smooth Curve Rotating in Depth.** The curve of Figure 4.10 is rotated 87° from its initial position, and then undergoes an instantaneous rotation about the central vertical axis of the cylinder. (a) The two-dimensional projection of the curve and its true velocity field. (b) The initial perpendicular components of velocity. (c) The computed velocity field of least variation.

rotation. The curves are rotated 70° and 87°, respectively, from the initial position shown in Figure 4.10. As in the case of the circle rotating in depth, the smoothest velocity fields for this curve are similar to the true velocity fields, but deviate from the true directions as distance from the central horizontal axis of the curve increases. This deviation is present at the three viewing angles shown here.

In Figures 4.13, 4.14 and 4.15, this same three-dimensional curve is rotated about the axis $\mathbf{n} = [1,1,1]^T$. If we let the position of the curve in Figure 4.13 be the zero position, then the curves in Figures 4.14 and 4.15 occur at a rotation of 90° and 180°, respectively, from this initial position. Again, the true projected velocity field,

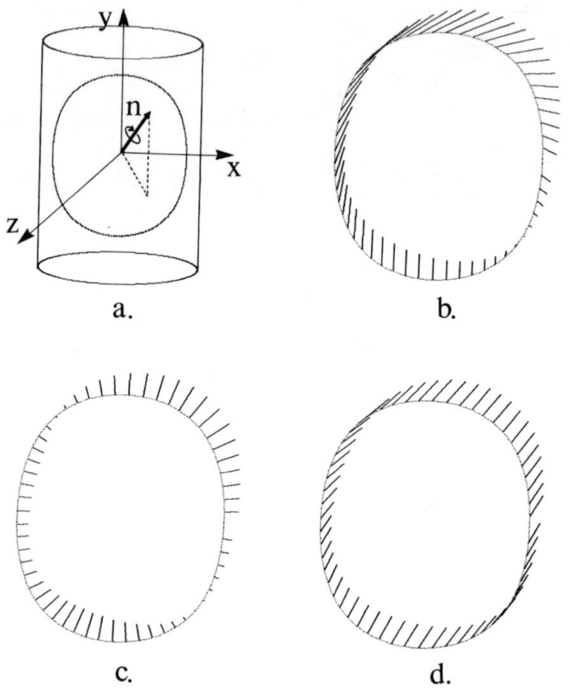


**Figure 4.13. Smooth Curve Rotating in Depth.** (a) A circle, wrapped around an imaginary cylinder, rotating around the axis $\mathbf{n} = [1, 1, 1]^T$. (b) The two-dimensional projection of the curve and its true velocity field. (c) The initial perpendicular components of velocity. (d) The computed velocity field of least variation.

initial perpendicular velocity components, and computed smoothest velocity fields are shown, for each of the three positions of the curve. At the initial position of the curve, shown in Figure 4.13, the smoothest velocity field is significantly different from the true velocity field, exhibiting a large translational component toward the upper right corner of the image. At other positions of the curve, such as those in Figures 4.14 and 4.15, the true and smoothest velocity fields are qualitatively very similar. A subsequent computation of three-dimensional structure from the smoothest velocity field cannot possibly derive the correct structure of this curve on the basis of its motion around the initial position shown in Figure 4.13. The similarity of the true and computed velocity fields at other positions of the curve, such as those in Figures 4.14 and 4.15, may yield a three-dimensional interpretation that is much closer to the true structure of the curve.

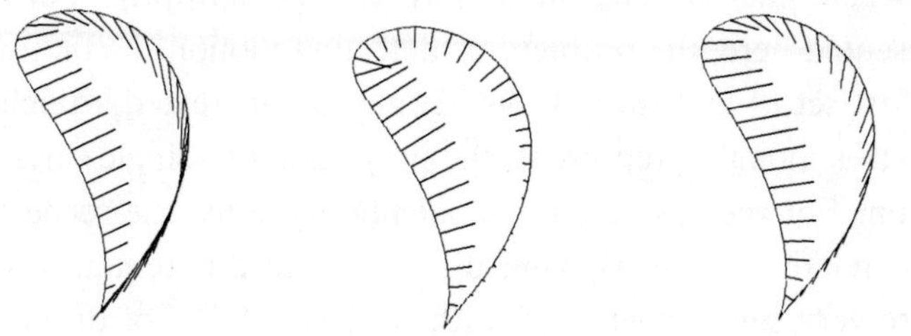

a. b. c.

**Figure 4.14. Smooth Curve Rotating in Depth.** The curve of Figure 4.13 is rotated 90° from its initial position, and then undergoes an instantaneous rotation about the axis $\mathbf{n} = [1, 1, 1]^T$. (a) The two-dimensional projection of the curve and its true velocity field. (b) The initial perpendicular components of velocity. (c) The computed velocity field of least variation.

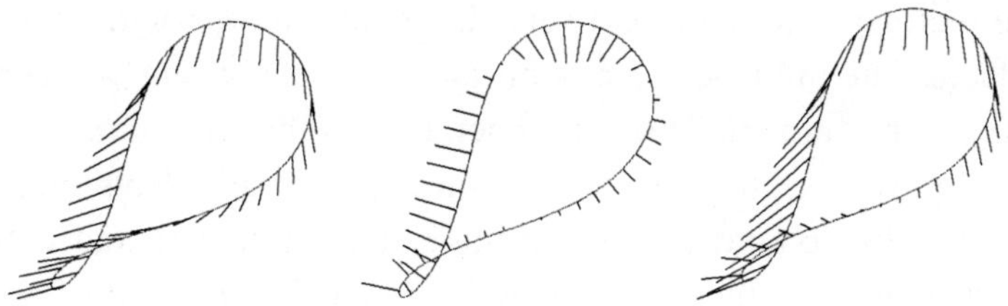

a. b. c.

**Figure 4.15. Smooth Curve Rotating in Depth.** The curve of Figure 4.13 is rotated 180° from its initial position, and then undergoes an instantaneous rotation about the axis $\mathbf{n} = [1, 1, 1]^T$. (a) The two-dimensional projection of the curve and its true velocity field. (b) The initial perpendicular components of velocity. (c) The computed velocity field of least variation.

### 4.4.2. Natural Image Sequences

The final set of examples includes contours that have been extracted from natural images. For the first set of examples, the motion was constructed artificially. That is, a single natural image was translated or rotated to produce a second image. In this case, the known velocity field allows a rigorous evaluation of the correctness of the resulting solution. The final example is a sequence of aerial photographs. Here, we rely on a qualitative assessment of the results.

In general, a natural scene contains multiple objects that may undergo different motions. Even in a static scene, the movement of the observer creates differential motion at object boundaries. The smoothness constraint is not valid at object boundaries, where the velocity field may change abruptly. For the particular examples presented here, the problem of detecting motion discontinuities does not arise. In the first set of examples, in which motion is created artificially, the entire image is treated as though it represents the projection of a single surface, undergoing a single motion. For the case of the aerial photographs, the scene is essentially a single surface, relative to the viewing distance. The differential motions at object boundaries are very small, compared with the overall size of displacements in the image. In general, the detection of object boundaries, either from motion cues or other visual cues, is important for a velocity field computation that utilizes the smoothness constraint. It should be stressed, however, that the smoothness constraint is only imposed over single image contours, or parts of single contours (an issue that is addressed further in Section 4.5.1). In a highly textured image, analyzed at a fine resolution, zero-crossing contours are very dense, and most contours are confined to single surfaces. The influence of motion measurements along one contour does not extend to other contours. Thus, in general, smoothness is imposed only over a very localized portion of the image, which reduces the problems that arise from the presence of object boundaries. In addition, this limited spatial extent of the smoothness constraint is useful for the analysis of motion generated by scenes such as waterfalls, or trees blowing in the wind, in which local surface structure, and the projected velocity field, changes vary rapidly over space and time.

**1. Pure translation of arbitrary curves from a natural image**

The image of Figure 4.16a was translated to the right by three picture elements, to yield a second image. The two images were convolved with a $\nabla^2 G$ operator, whose central positive diameter was 12 picture elements, and the zero-crossings, shown in Figure 4.16b, were obtained. The time derivative, $\frac{\partial}{\partial t}(\nabla^2 G * I)$ was computed by subtracting the two filtered images. At the location of a zero-crossing, the local orientation of the perpendicular vector $\mathbf{u}^{\perp}$ was computed from the gradient of the filtered image. The perpendicular components of velocity, $v^{\perp}$, were then computed as follows:

$$v^{\perp} = \frac{-\frac{\partial}{\partial t}(\mathbf{I}')}{|\nabla \mathbf{I}'|}$$

where $\nabla$ denotes the gradient operator, and $\mathbf{I}'$ is the filtered image. When discrete images are used to obtain the initial motion measurements, there is a limit on the

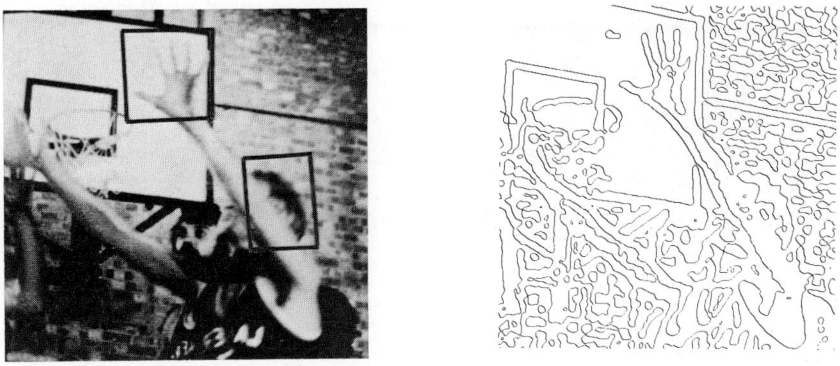

a. b.

**Figure 4.16. A Natural Image.** (a) The original image, which was either translated or rotated to produce a second image. The image contains 576 × 576 picture elements. The rectangles outline areas where the contours of Figures 4.17 and 4.18 have been extracted. (b) The zero-crossing contours, derived from the convolution of the image with a $\nabla^2 G$ operator.

spatial displacement of features that can be tolerated. The reason for this can be seen from the illustration of the Marr-Ullman model in Figure 2.4. Figure 2.4a showed the response of the one-dimensional $D^2G$ operator to a step change of intensity. If we let $w_x$ denote the width of the central positive region of the operator, then the distance between the positive and negative peaks of the profile in Figure 2.4a is $w_x$. In the case of an ideal, isolated intensity change, a displacement of the intensity profile up to about $w_x$ results in a temporal derivative that yields the correct sign of $v^\perp$ at the zero-crossing. The reliable measurement of the magnitude of $v^\perp$ requires a displacement of at most $\frac{w_x}{2}$. In a highly textured image, however, displacements as small as $\frac{w_x}{2}$ can still yield significant error in the sign or magnitude of $v^\perp$.

The analysis of the velocity field is presented for zero-crossing contours extracted from the areas outlined by the rectangles in Figure 4.16a. The contours, together with their true velocity field, are shown in Figures 4.17a and 4.18a. The initial perpendicular velocities are shown in Figures 4.17b and 4.18b. In Figure 4.19, the perpendicular vectors $v^\perp \mathbf{u}^\perp$ along the contours are displayed as points in velocity space. Measurements derived from the curves in Figures 4.17 and 4.18 are displayed in Figures 4.19a and b, respectively. If the initial measurements were perfect, the points in velocity space would lie along the solid circles shown. The deviation of the position of the points from this circle provides a visual demonstration of the large

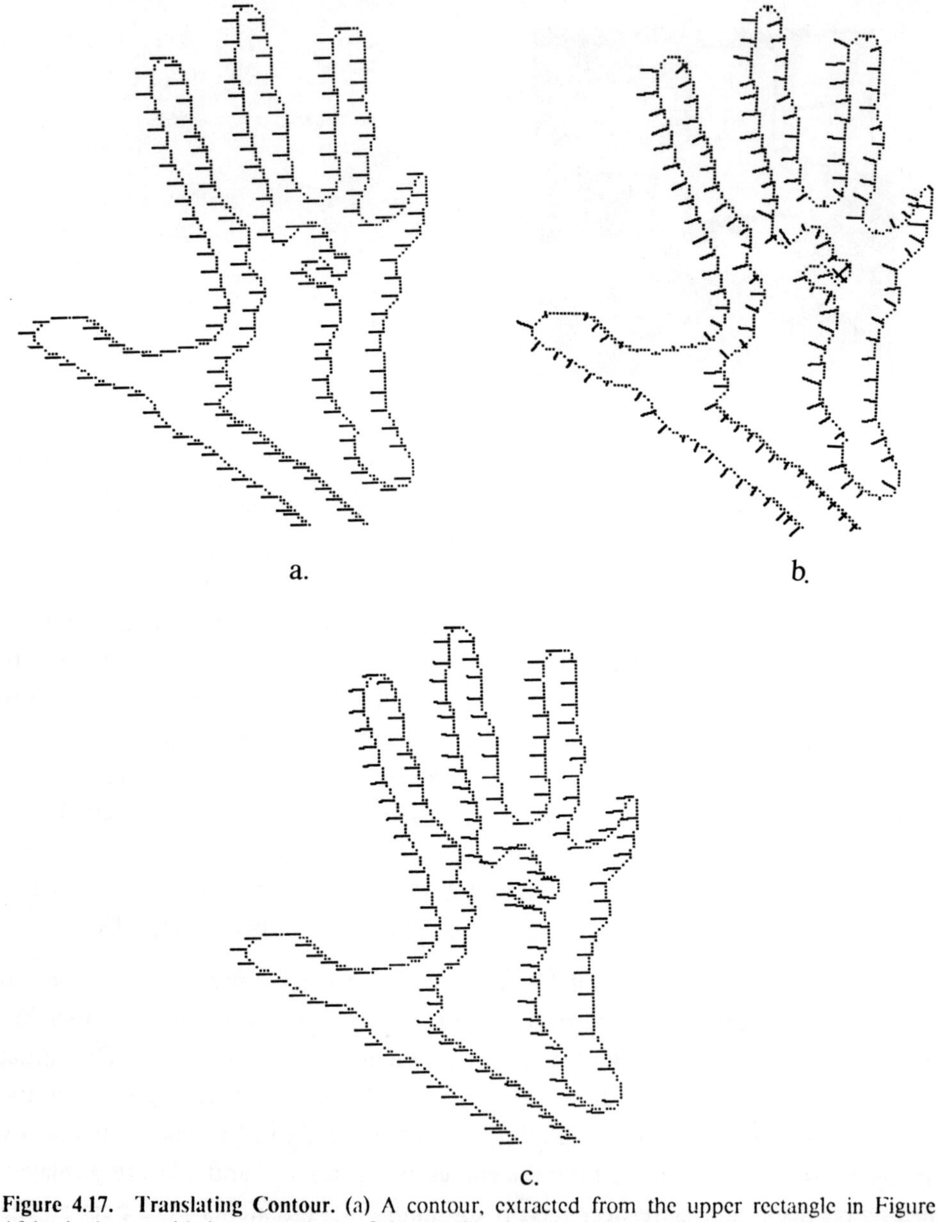

**Figure 4.17. Translating Contour.** (a) A contour, extracted from the upper rectangle in Figure 4.16a, is shown with its true velocity field. (b) The initial perpendicular components of velocity. (c) The computed velocity field of least variation.

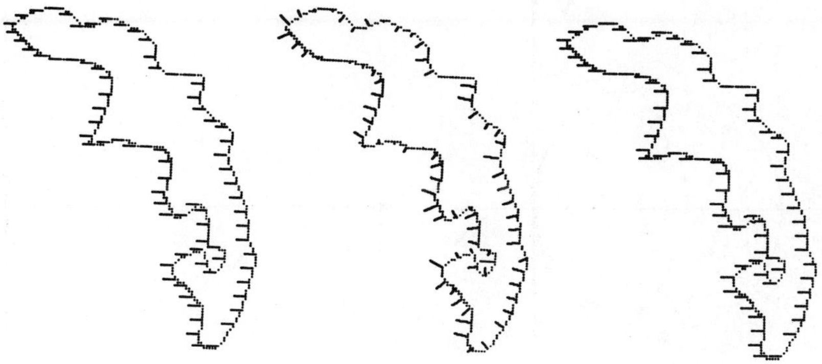


Figure 4.18. **Translating Contour.** (a) A contour, extracted from the lower rectangle in Figure 4.16a, is shown with its true velocity field. (b) The initial perpendicular components of velocity. (b) The computed velocity field of least variation.

error in the initial motion measurements. The resulting velocity field solutions, with $\beta = 0.001$, are shown in Figures 4.17c and 4.18c. For the computed velocity field of Figure 4.17c, the average error in the direction of velocity is 2.2°, and average error in magnitude is 2.8% of the true magnitude of velocity. The directions of all resulting velocity vectors are within 5° of the true direction of velocity. For the contour of Figure 4.18, the average error in computed directions is 0.85°, and average error in the magnitudes of velocity is 3.7%. In this case, all directions of velocity are within 3° of the true direction.

It is significant that a velocity field solution can be obtained, for which the error in velocity is so small, given the large error in the perpendicular components of velocity that formed the input to the algorithm. This error in the input underscores the need to design an algorithm that only requires the velocity field to satisfy the image constraints approximately.

**2. Rotation of arbitrary curves from a natural image**

For the next two examples, the original image was rotated rigidly to obtain a second image. Figure 4.20a shows the zero-crossing contour that is derived from the upper rectangle of Figure 4.16, and its true velocity field. The original $\nabla^2 G$ operator had a central positive diameter of 16 picture elements. The image was rotated 3°

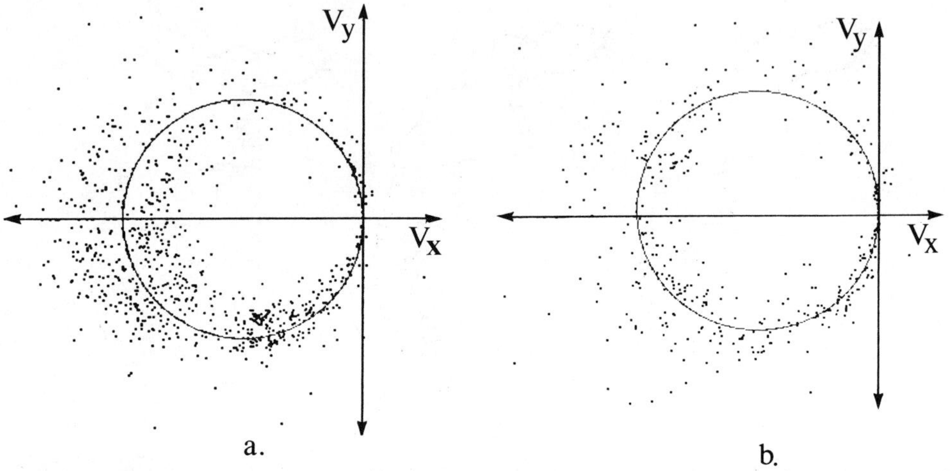


**Figure 4.19. Constraints in Velocity Space.** (a) The vectors $v^{\perp}\mathbf{u}^{\perp}$, measured along the contour of Figure 4.17, are indicated by points in velocity space. Perfect measurements would lie along the circle shown. (b) The same measurements, for the contour shown in Figure 4.18.

in the clockwise direction, about a point in the center of the palm of the hand. The length of the displacement vectors in Figure 4.20 is slightly exaggerated. Again, the initial perpendicular components of velocity were computed from the time derivative between the two filtered images, and the gradient along the zero-crossing contours of the first filtered image; these components are illustrated in Figure 4.20b. The computed smoothest velocity field, with $\beta = 0.001$, is shown in Figure 4.20c. The true and computed velocity fields are shown superimposed in Figure 4.20d. The agreement between the directions of velocity for the true and computed motions is quite good around the fingers, thumb, and rightmost boundary of the hand. Around the palm of the hand, and the contour running from the thumb to the forefinger, there is considerable error in the direction of velocity. Error in the area of the palm of the hand may be due to a reduced contrast along the contour. The average error in the direction of velocity over the entire contour is 6.1°. The magnitudes of velocity are generally smaller for the smoothest velocity field; average error in magnitude is 23.1% of the true magnitude of velocity. The solution varies with the choice of $\beta$; a larger value of $\beta$ leads to better agreement in the magnitudes of

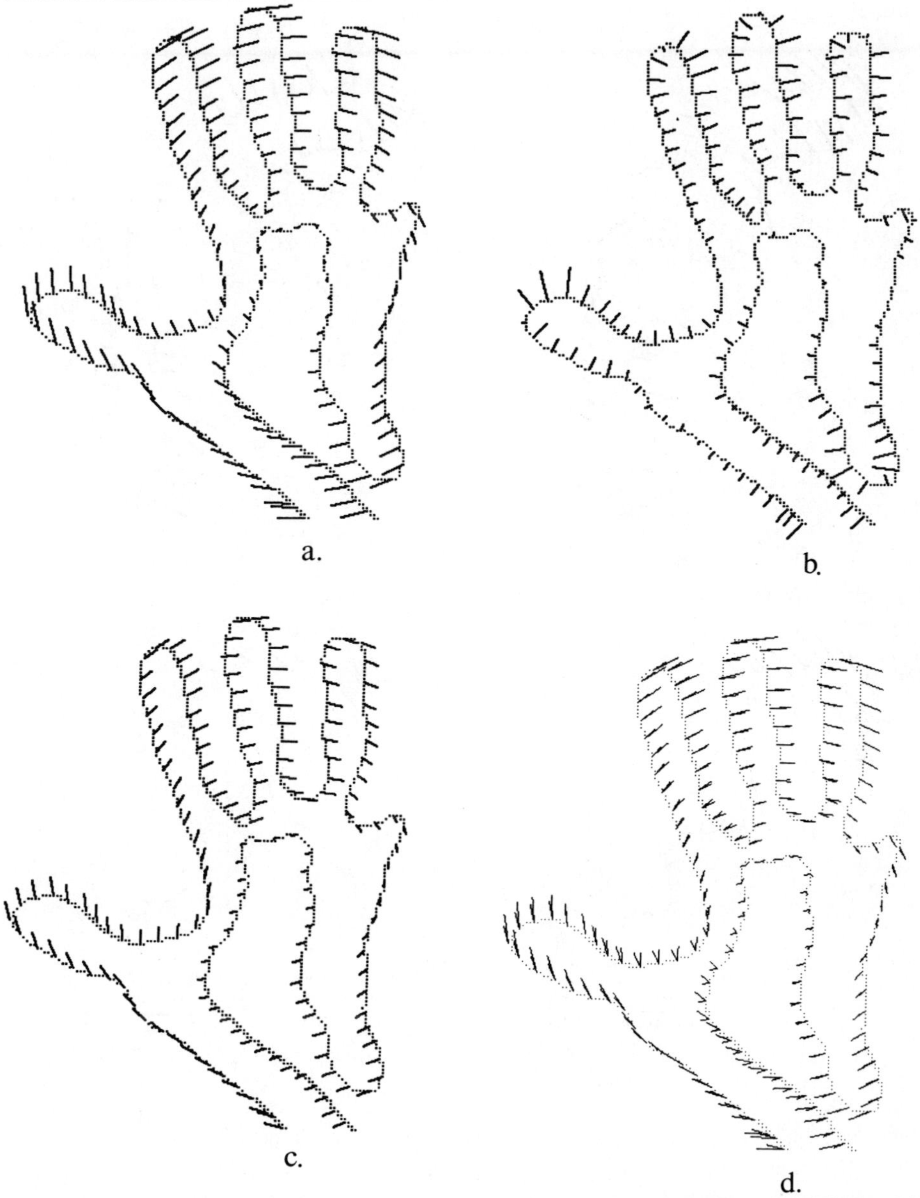

**Figure 4.20. Rotating Contour.** (a) A contour, extracted from the upper rectangle in Figure 4.16a, is shown with its true velocity field. (b) The initial perpendicular components of velocity. (c) The computed velocity field of least variation. (d) The true and computed velocity fields superimposed.

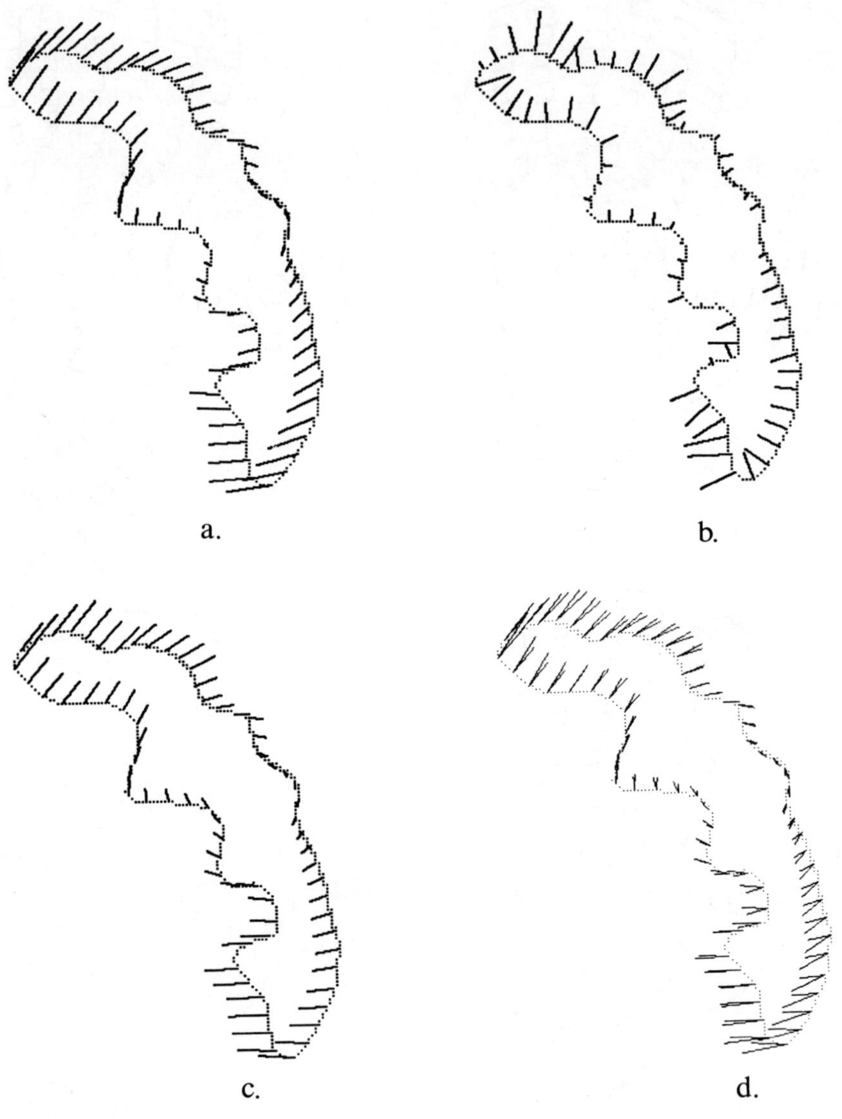

Figure 4.21. Rotating Contour. (a) A contour, extracted from the lower rectangle in Figure 4.16a, is shown with its true velocity field. (b) The initial perpendicular components of velocity. (c) The computed velocity field of least variation. (d) The true and computed velocity fields superimposed.

velocity between the true and smoothest velocity fields, but the error in direction of velocity increases.

A second example is shown in Figure 4.21. This zero-crossing contour was derived from the lower rectangle of Figure 4.16, again using a $\nabla^2 G$ operator whose central positive diameter was 16 picture elements. In this case, the image was rotated rigidly about a point in the center of this contour. The contour, with its true velocity field, is shown in Figure 4.21a. The initial perpendicular components of velocity are shown in Figure 4.21b, and the computed smoothest velocity field appears in Figure 4.21c. Figure 4.21d shows the true and smoothest velocity fields superimposed. In this case, $\beta = 0.01$. Again, the two velocity fields are qualitatively similar, but differ quantitatively. Average error in the magnitude of velocity is 13.7% of the true magnitude, and average error in the direction of velocity is 10.3°.

In general, the velocity field of least variation, for smooth curves in rotation, is expected to differ from the true velocity field. This will be the case for rotation in the image, as well as rotation in space, and was demonstrated for a number of ideal smooth curves in the previous section. The error in the velocity fields of Figures 4.20 and 4.21 is due in part to error in the initial motion measurements, and in part to the fact that even with ideal input, we do not expect to obtain the correct velocity field in this case.

### 3. A natural motion sequence

Figures 4.22a and b show two aerial photographs, taken in sequence from an airplane. The images contain $256 \times 256$ picture elements. The two images were convolved with a $\nabla^2 G$ operator whose central diameter was 12 picture elements. The resulting zero-crossings, for the two images, are shown in Figures 4.23a and b. The contours cover a large extent of the image in this case; this is a consequence of the small size of the image, and large displacements, which required a large initial $\nabla^2 G$ operator. If the image were analyzed with higher spatial and temporal resolution, the two-dimensional velocity field could also be obtained with higher resolution. In Figure 4.24, the two zero-crossing descriptions are shown superimposed, in order to illustrate the relative displacements between the two images. The zero-crossings from the first image are shown in white, those from the second are black, and the background is grey. Points at which zero-crossings occurred in both images are shown in white. Qualitatively, it can be seen that displacement increases in

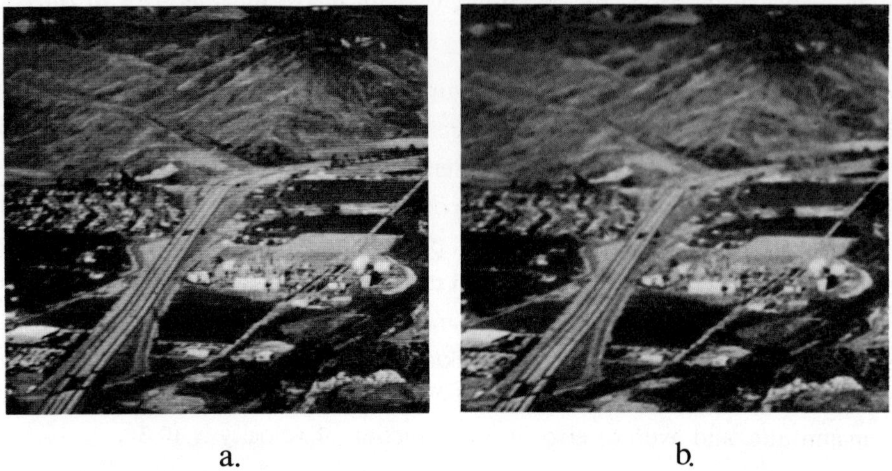


**Figure 4.22. A Natural Image Sequence.** (a) and (b) represent two images in a sequence, taken from an airplane. The images contain 256 × 256 picture elements.

magnitude as we move from the top to the bottom of the image. In addition, the displacements vary in direction over the image, having a larger horizontal component toward the top of the image, and a larger vertical component toward the bottom.

As before, the initial perpendicular components of velocity were computed along the zero-crossing contours in the first filtered image by comparing the local gradient at the zero-crossings with the time derivative, obtained by subtracting the two filtered images. The algorithm was then run over each contour separately, to obtain the velocity field along the zero-crossing contours of the first image. In Figure 4.25, the zero-crossings are shown in black, with the resulting velocity vectors in white. The vectors shown represent a sampling of the velocity field along the contours. Very small contours, and some of the contours that occurred at the image boundary, were not included in this analysis. The length of the vectors in this display is equal to their displacement in the image. Qualitatively, the magnitude of the displacement vectors increases from the top to the bottom of the image, and their directions also vary in a way that reflects the displacement of the zero-crossing contours in Figure 4.24.

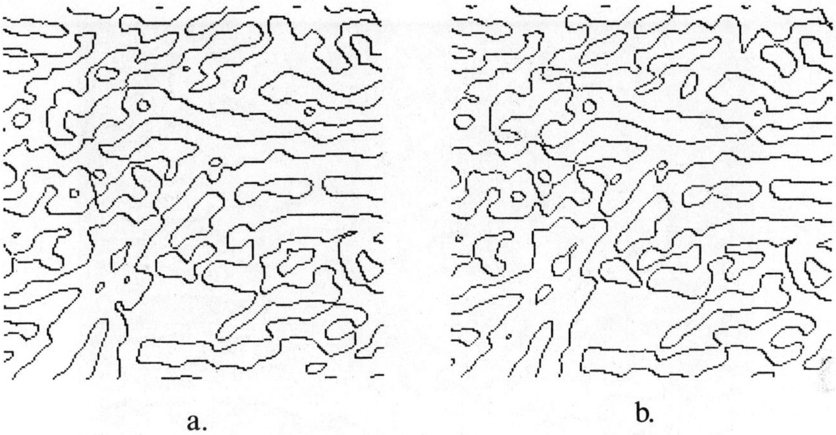


Figure 4.23. Initial Zero-crossing Descriptions. (a) and (b) illustrate the zero-crossings that are derived from the convolution of the images of Figure 4.22a and b with a $\nabla^2 G$ operator whose central diameter was 12 picture elements.

In Figure 4.26, the first image, reduced in contrast, is shown with the velocity field superimposed with black lines. The velocity vectors are evenly spaced on the image, and were computed by taking the average of the velocity vectors within a neighborhood of size $48 \times 48$ picture elements, centered at each point. The length of the velocity vectors has been doubled in this display, in order to emphasize the variation in direction and magnitide of velocity over the image. The resulting velocity field agrees qualitatively with the displacements of the contours shown in Figure 4.24. It also agrees well with the perceived movement between the images, when viewed in rapid succession.

### 4.4.3. Summary of Experimental Results

The velocity field algorithm has been tested on two classes of motion input: the first consisted of ideal smooth curves, whose perpendicular components of velocity were computed analytically; the second consisted of natural image sequences. The first class of examples allowed the examination of the velocity field of least variation, given the best possible input. The second class provided a test of the robustness of the approach in more practical imaging situations.

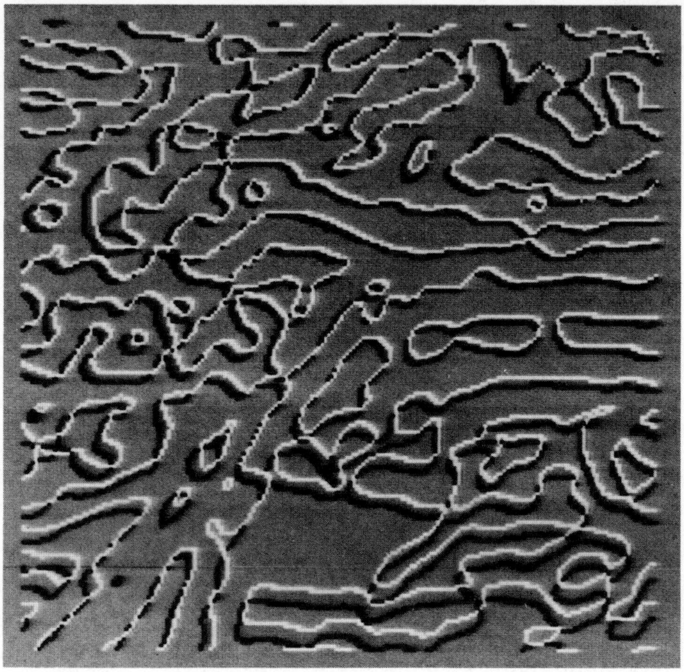

**Figure 4.24. Superimposed Zero-crossings.** The zero-crossings of Figure 4.23a are shown in white, and those of Figure 4.23b are black. Points that give rise to zero-crossings in both images are also shown in white.

With regard to ideal curves, a set of examples was first presented, for which it had been shown theoretically that the smoothest velocity field is the correct one; these examples included the translating polygon and ellipse, the rotating polygon, and the Kinetic Depth Effect stimulus. The algorithm was then tested with a number of smooth curves in rotation. For these examples, we lacked theoretical results concerning the correctness of the velocity field solution, and were relying on an empirical investigation. In many cases, the velocity field of least variation was qualitatively similar to the true velocity field; for example, in the cases of the circle rotating in depth, and the smooth curve of Figure 4.10 rotating around the vertical axis. Even fairly complicated velocity fields, such as those in Figures 4.14 and 4.15, were matched quite well by the smoothest velocity field. In some examples, such as the curve underlying the barberpole illusion, and the rotating curve of Figure 4.13,

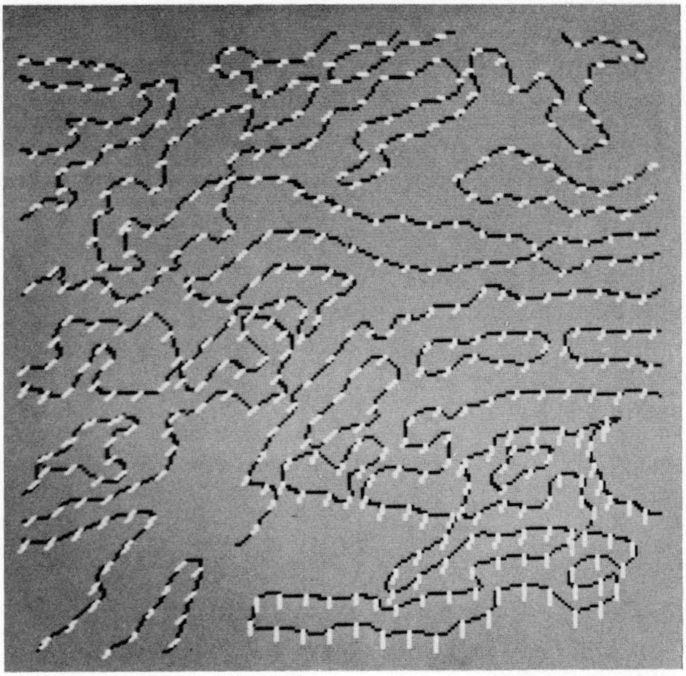

**Figure 4.25. The Computed Velocity Field.** The zero-crossing contours are shown in black. A sampling of the computed velocity vectors along the contours is shown in white. The magnitudes of the vectors are equal to the displacement of the contours between the two images.

the true and computed velocity fields were very different. It does not appear from these examples, that the correctness of the velocity field depends on an obvious property of the image curves, such as their curvature. The same curve in Figures 4.10 and 4.13 yields a velocity field very similar to the true motion of the curve when rotated about the vertical axis, but a very different velocity field when rotated about the axis $\mathbf{n} = [1, 1, 1]^T$.

The algorithm performed reasonably well on the natural image sequences. For the translating contours of Figures 4.17 and 4.18, the error in the final solution was very small, relative to the large error in the perpendicular velocity components that formed the input to the algorithm. The error increased for the rotating curves of Figures 4.20 and 4.21, although the computed motion was still similar to the true

**Figure 4.26. The Computed Velocity Field.** At evenly spaced points over the image, the average velocity was computed for a patch of size 48 × 48 picture elements around the points. The average velocity vectors are displayed in black.

motion of the curves. For the final natural motion sequence, the computed velocity field agreed well with the apparent displacement of the zero-crossing contours between the two frames.

## 4.5. The Convergence of Iterative Methods

A disadvantage of iterative methods, in general, is that a large number of iterations is typically required for convergence to the final solution. The conjugate gradient algorithm exhibits relatively fast convergence, compared with other methods. If there are $n$ parameters to compute, this algorithm is guaranteed to converge in at most $n$ iterations. The actual convergence depends in part on the initial velocity field, and convergence is much faster for a closed contour. For the examples presented in the previous section, the curves generally consisted of several hundred points. For a curve of $n$ points, there are $2n$ parameters to compute (the $x$ and $y$ components of velocity). Thus, there were typically 800 — 1000 parameters. In the case of ideal smooth curves, convergence required from 80 to 200 iterations, depending on the similarity between the initial and final velocity fields, and whether the curve was open or closed. For the contours extracted from natural images, the algorithm only approached a final solution asymptotically, due to the large error in the initial measurements, and numerical error in the computation. A curve of $n$ points typically required $n$ iterations to bring the solution within 5% of the final solution, and another $\frac{n}{2}$ to $n$ iterations to reach a stage at which the change in parameters from one iteration to the next was less than 0.01%.

In this section, we discuss three methods for improving the convergence of an iterative method for the motion computation: (1) piecewise optimization; that is, the contour is divided into parts, and the minimization applied to each part in parallel, (2) the use of multiple frames in a motion sequence, with fewer iterations per frame, and (3) the use of multi-grid techniques (Brandt 1977a, 1977b).

### 4.5.1. Piecewise Optimization

For the iterative methods discussed in this chapter, the rate of convergence depends on the number of points on the contour. The number of iterations required can be reduced if the curve is divided into parts, and the minimization performed on each part in parallel. This raises the problem of how to divide the curve. One issue that must be considered is the range of orientations spanned by a segment of the curve. It was shown in Section 3.3.2 that at least two measurements of $v^{\perp}(s)$ from points of different orientation on the contour, are required to guarantee a unique velocity field solution. Reliability of the solution requires measurements at orientations that differ significantly; the reason is illustrated in Figure 4.27. Suppose an object is translating in the image plane with velocity $\mathbf{V}_o$, and that the constraint imposed by two measurements of $v^{\perp}(s)$, from points that differ in orientation by only $10°$, gives rise to the two lines in velocity space labelled $l_1$ and $l_2$ in Figure 4.27. If the measurements were perfect, these lines would intersect at the endpoint of the vector $\mathbf{V}_o$. In practice, there will be error in these measurements. If the actual measurements differ from the true measurements by as much as 10%, the lines of constraint in velocity space would fall somewhere within the grey bands

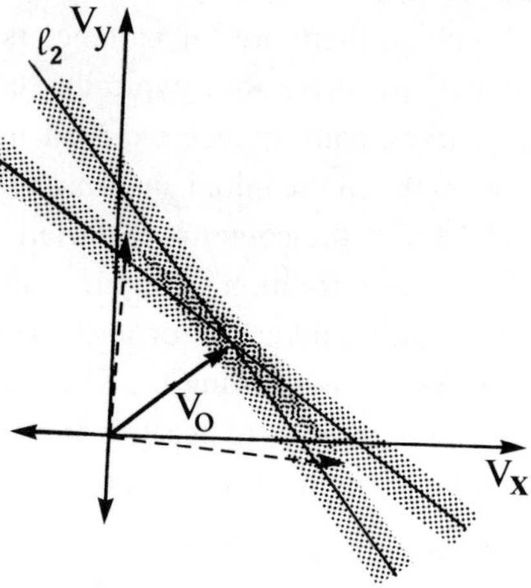


**Figure 4.27. The Consequence of Constraint Error.** The lines of constraint, $l_1$ and $l_2$, are derived from a curve that is translating with velocity $V_o$, and differ in orientation by only 10°. An error of 10% in the magnitudes of $v^{\perp}$ can result in an intersection of the constraint lines that falls anywhere within the dotted arrows.

surrounding $l_1$ and $l_2$ in the figure. The intersection of these constraint lines could imply a direction of velocity of the object that differs significantly from its true direction, as indicated by the dotted arrows. As a consequence, a larger range in the orientations spanned by a segment of the curve, from which the measurements of $v^{\perp}(s)$ are derived, leads to greater reliability of the solution. This need to combine velocity constraints over a significant range of orientations is fundamental to the motion measurement computation. Schemes that combine constraints in velocity space, assuming pure translation (Fennema and Thompson 1979; Marr and Ullman 1981; Thompson and Barnard 1981; Adelson and Movshon 1982; Schunck 1983), also face this problem.

There are several ways in which a curve can be divided into parts. One strategy is to divide the curve into segments that span a particular range of orientations. The actual range depends on the expected reliability of the measurements. To use this strategy, we need to take into consideration the fact that error in $v^{\perp}(s)$ is generally greater at areas of high curvature on the contour. A small segment of the curve that has very high curvature, may not yield a reliable velocity measurement. An extended straight line does not provide measurements at the range of orientations that are necessary, but generally yields more reliable measurements of $v^{\perp}(s)$. A second possible strategy is to split the curve at maxima or minima of curvature, as suggested by Hoffman and Richards for the representation of curves for recognition (Hoffman and Richards 1982). If features, such as points of high curvature, could be identified unambiguously in two adjacent frames, they could be tracked over time, and used to divide the curve.

After the curve is divided, the minimization may be performed on each part in parallel. Once solutions are obtained for the parts, they can be united again, and a minimization performed over the entire curve, in order to obtain continuity of the velocity field at the boundaries between parts. This final stage should converge very fast, because the solutions obtained for the parts should be close to the final solution.

### 4.5.2. Multiple Frames

While it is possible to compute the full two-dimensional velocity field on the basis of a single curve and the instantaneous measurements of $v^{\perp}(s)$, the computation could instead be extended over a larger time interval. At each moment, only a small number of iterations may be performed, and the current velocity field could then be passed on to the curve at the next moment in time. An extended period of time, or several frames, may be required before the velocity field converges to a solution. Horn and Schunck (1981) discuss the trade-off between the number of frames, and number of iterations per frame, for the optical flow computation.

### 4.5.3. Multi-grid Methods

A primary reason for the slow convergence of the iterative methods discussed in this chapter is the global propagation of information via local operations. Recently, a class of iterative techniques called multi-grid methods have been used in the numerical solution of boundary value problems, for which they achieve a significant improvement in convergence over previous iterative methods (Brandt 1977a, 1977b). These techniques take advantage of a hierarchy of resolutions, in order to increase the efficiency of global propagation of information. Interactions between computations at different resolutions exploit information at coarser levels for the computations at finer levels of resolution. The application of these techniques to the visual processing problem of surface reconstruction from sparse stereo data has led to an order of magnitude decrease in the number of iterations required for the computation, in comparison with other iterative methods (Grimson 1981; Terzopoulos 1982). For the motion computation, multiple resolutions could be obtained by sampling the curve at different resolutions, or by making use of information supplied by multiple spatial frequency filters, as in Terzopoulos' work on surface reconstruction.

### 4.6. Summary

In this chapter, possible algorithms for computing the velocity field along contours were described, which use the additional smoothness constraint to determine the velocity field uniquely. The computation may be formulated as an optimization problem, whose solution is equivalent to the solution of a system of linear equations. To solve the problem, standard iterative algorithms were considered. For the case in which constraints from the image are to be satisfied exactly, a contour of $n$ points gives rise to a system of $n$ linear equations in $n$ unknowns. The application of Gauss-Seidel and Jacobi relaxation to the solution of this system of equations was illustrated. For the case in which the constraints are to be satisfied only approximately, the solution to the resulting optimization problem is equivalent to the solution of a system of $2n$ linear equations in $2n$ unknowns, given by the $x$ and $y$ components of velocity. To compute the velocity field in this case, we utilized a standard algorithm from mathematical programming, known as the conjugate gradient algorithm.

In general, the nature of the computation is such that it leads naturally to algorithms that involve simple, local, parallel operations. In this sense, it is a feasible computation to perform, for a biological system. The empirical results of the algorithm, for ideal smooth curves, as well as curves from natural images, also support the physical plausibility of the computation. Finally, we show in Chapter 6 that the predictions of the theory, for the smooth curves in rotation presented in Section 4.4.1, are qualitatively consistent with human motion perception.

# 5. THE COMPUTATION OF MOTION DISCONTINUITIES

If two adjacent surfaces undergo different motions, they generally give rise to a discontinuity in the velocity field along their boundary. Across such a boundary, the physical constraint of surface smoothness, utilized in the velocity field computation, is not valid. If image contours (such as the zero-crossing contours) were confined strictly to single surfaces, then a contour-based velocity field computation could restrict the combination of initial motion measurements to those derived from the same physical surface. Some surface boundaries are associated with sharp reflectance changes that give rise to zero-crossing contours that closely follow the boundary. In these situations, the contours are usually confined to single surfaces. In the image of the basketball game in Figure 4.16, for example, the boundaries of the player's hand against the backboard, and of the backboard itself, give rise to zero-crossing contours that roughly follow the boundaries. In general, however, zero-crossing contours may cross surface boundaries, particularly in highly textured images. The integration of initial motion measurements along these contours leads to error in the velocity field solution at the boundaries. It would be advantageous to detect these boundaries prior to the velocity field computation, so that initial motion measurements are combined only along parts of contours that arise from the same physical surface.

There are many visual sources of information about the location of surface boundaries; for example, sharp changes in depth, motion or texture may reveal the location of a boundary. Any of these sources can, in principle, be used to establish boundaries for the velocity field computation. Perceptual studies suggest that the human visual system may be capable of identifying boundaries on the basis of motion information alone. A movie consisting of two random dot patterns, in which a central patch in the second frame is shifted relative to the first (see Figure 1.4), elicits a vivid impression of a central object separated from its background, with sharply defined boundaries (for example, Anstis 1970; Regan and Spekreijse 1970; Julesz 1971; Braddick 1973, 1974). This suggests that the use of relative movement to establish surface boundaries can proceed in the absence of all other cues to the organization of the scene. The real world provides natural examples of our ability to isolate objects on the basis of motion information alone, through the use of camouflage. There are butterflies, for example, whose colors allow them to blend naturally into the bark of a tree; if the butterfly is very still, it remains unnoticed, but should it move only slightly, we suddenly detect its presence.

In this chapter, we begin to examine the computational problem of detecting surface boundaries from relative movement, independent of other sources of boundary information. While the detection and localization of surface boundaries in natural scenes may ultimately involve the integration of information from many visual sources, it is useful to know what can be computed from motion information alone. The main conclusion of this chapter is that the detection of motion discontinuities can begin to take place prior to the combination of initial motion measurements for the full velocity field computation.

In Section 5.1, related work on the detection of motion discontinuities is discussed. In the remainder of the chapter, we then focus on the design of an algorithm to compute a rough description of discontinuities from the initial measurements of $v^{\perp}(s)$ along zero-crossing contours. We first discuss some of the inherent limitations of such an approach, and show that this detection is more difficult when surfaces are allowed to rotate as well as translate. One simple detection algorithm is presented, with results of the application of the algorithm to a complex dynamic random dot pattern. The algorithm has been tested only with dense random dot patterns, and performs well in this limited domain. Much further work is required, however, to develop a robust algorithm that also performs well for natural image sequences. The purpose of the work presented here is only to establish the feasibility of detecting motion discontinuities, prior to the full velocity field computation.

## 5.1. Related Work on Discontinuity Detection

Previous work on the computation of motion discontinuities has utilized two basic approaches. In the first, the velocity field is computed, and then followed by a stage of analysis that locates significant changes of velocity in the image (for example, Potter 1975, 1977; Thompson 1980; Thompson, Mutch and Berzins 1982; Mutch and Thompson 1982). In the second approach, discontinuities are inferred prior to the velocity field computation (for example, Schunck and Horn 1981). In the work of Thompson, Mutch and Berzins (1982; Mutch and Thompson 1982), the velocity field is computed using a local correspondence scheme that exploits displacement information at a range of image resolutions. Sparse velocity vectors are interpolated by blurring, to produce a dense velocity field. Discontinuities are then detected in a manner similar to the detection of intensity changes by Marr and Hildreth (1980). That is, discontinuities are defined as zero-crossings in the Laplacian of either the $x$ or $y$ components of velocity. A range of image resolutions is used in this discontinuity detection. Thompson (1980) and Potter (1975, 1977) also compute the velocity field first; Thompson utilizes the gradient scheme described by Fennema and Thompson (1979), and Potter utilizes a correspondence scheme (described briefly in Section 3.1). Rather than explicitly detecting motion discontinuities, Thompson and Potter utilize region growing techniques to group elements of similar velocities. Discontinuities are then given implicitly by the boundaries between regions. These latter two schemes are presently limited to the segmentation of scenes consisting only of objects undergoing pure translation, while the previous schemes, based on locating discontinuities in the $x$ or $y$ components of velocity, allow surfaces to rotate as well (although the initial computation of the velocity field assumes pure translation).

A difficulty with the first approach described above, is that without prior knowledge of surface boundaries, it is difficult to compute the velocity field correctly. The need to combine local motion constraints over an area of the image, or along contours, often leads to error in the computation of the velocity field at surface boundaries. This problem led Schunck and Horn (1981) to consider methods for inferring boundaries from the initial measurements of the components of motion provided in gradient schemes. They first observed that the combination of motion constraints, imposed by two adjacent surfaces undergoing different motions, often yields a local velocity vector with extremely large magnitude, which is physically implausible. Second, they observed that the directions of the intensity gradients (and resulting lines of constraint in velocity space, illustrated in Figure 3.1) sometimes differ for points on adjacent surfaces. These observations were transformed into two heuristics for deciding when local motion constraints should be combined. If either (1) the orientations of two lines of constraint on local velocity, derived from nearby points in the image, differ by more than some angle $\phi$, or (2) the lines of constraint intersect in velocity space at a point whose distance from the origin is larger than some distance $\rho$, then the two motion constraints are assumed to arise from points on different surfaces. In this way, discontinuities can be inferred from properties of the initial motion measurements, prior to the computation of the full two-dimensional velocity field.[1]

Other work on the computation of motion discontinuities has been presented by Reichardt and Poggio (1979; Poggio, Reichardt and Hausen 1981; Reichardt, Poggio and Hausen 1983), who propose an algorithm for the detection of discontinuities in optical flow by the visual system of the housefly. Early processing mechanisms in the fly's visual system include both direction-selective movement detectors, and "flicker" detectors, which receive a single photoreceptor input, and signal changes in light intensity over time, at a particular visual location. Reichardt and Poggio argue that for the fly to evaluate relative movement, all flicker detectors must be inhibited, except those near discontinuities in the velocity field. They propose an algorithm in which a center-surround organization of directionally selective movement detectors is used to inhibit the flicker detectors when the same movement appears in the center and surround, so that significant activity of the flicker detectors indicates the position of motion discontinuities. Some of the basic ideas underlying this algorithm for the fly's visual system may extend to other visual systems as well.

## 5.2. The Early Detection of Discontinuities

In the remainder of this chapter methods are explored for inferring the location of discontinuities from properties of the initial perpendicular components of velocity, $v^{\perp}(s)$, along zero-crossing contours derived from the image. Advantages of this approach are first, the detection can be fast, because it takes place at the earliest possible stage in the measurement of motion, and second, it aids the next stage in the motion computation by establishing the boundaries within which local measurements are integrated, and smoothness imposed on the velocity field. A rough description of discontinuities, constructed from the initial motion measurements, could later be refined after the full velocity field computation.

[1]More recently, Schunck (1983) proposed an integrated approach to the detection of motion discontinuities. A rough estimate of the velocity field is computed initially, using a maximum likelihood estimation technique applied locally to the initial motion measurements provided by a gradient scheme. This computation is then followed by the integration of a procedure that detects motion discontinuities by finding changes in the $x$ or $y$ components of velocity, using standard edge detection techniques, and a procedure that smoothes the estimated velocity field. These latter two procedures provide significant improvement in the computed velocity field.

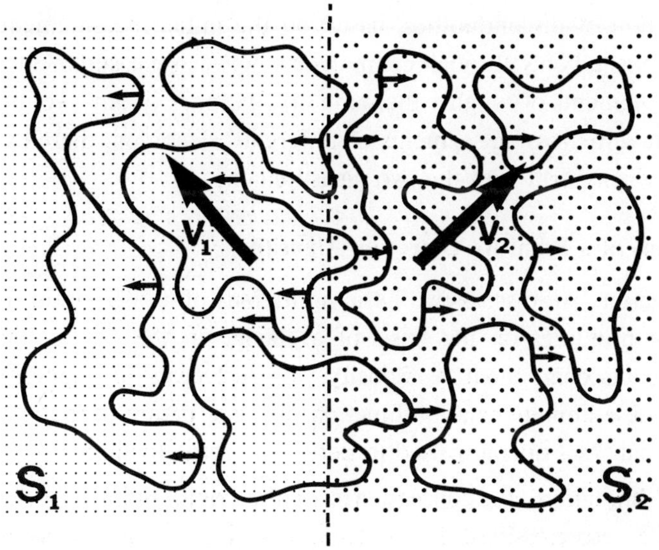


**Figure 5.1. Detecting Velocity Discontinuities.** The solid curves represent zero-crossing contours derived from the image of the two surfaces $S_1$ and $S_2$. The bold vectors $V_1$ and $V_2$ indicate the direction of velocity of the two surfaces. The smaller vectors indicate the direction of $v^{\perp}(s)\mathbf{u}^{\perp}(s)$ for vertically oriented segments of the zero-crossing contours; the directions of these vectors are different for the two surfaces.

The principle that makes it possible to detect discontinuities from the initial measurements of $v^{\perp}(s)$ along zero-crossing contours is illustrated in Figure 5.1. If two adjacent surfaces $S_1$ and $S_2$ undergo different velocities $\mathbf{V}_1$ and $\mathbf{V}_2$ (assuming pure translation for the moment), then zero-crossing contour segments whose orientation lies between the directions of $\mathbf{V}_1$ and $\mathbf{V}_2$ yield a change in the sign of $v^{\perp}(s)$ across the boundary. In Figure 5.1, for example, segments of the zero-crossing contours that are oriented vertically move to the left on $S_1$ and to the right on $S_2$. Thus, a change in the sign of $v^{\perp}(s)$, for zero-crossing segments of the same orientation, signals a discontinuity in the direction of velocity. In addition, zero-crossing segments whose orientation is such that the sign of $v^{\perp}(s)$ is the same on $S_1$ and $S_2$, may yield a change in the magnitude of $v^{\perp}(s)$ across the boundary. A discontinuity in the speed of movement alone could be detected only via a change in the magnitude of $v^{\perp}(s)$. Once the discontinuities are located, a contour that overlaps two surfaces can be divided, so that velocity constraints are only combined along parts of the contour that belong to a single surface. We should stress that the reliable detection of discontinuities, based on the initial measurements of $v^{\perp}(s)$, can be accomplished only through comparisons between measurements obtained at zero-crossing segments of roughly the same orientation. For any two measurements of $v^{\perp}(s)$ from zero-crossing segments of different orientation, there exists a single velocity vector consistent with the two measurements.

### 5.3. Limitations of a Primitive Detection Algorithm

In this section, we discuss two basic limitations of an algorithm that attempts to detect discontinuities on the basis of a change in the sign or magnitude of the perpendicular component of velocity, for zero-crossing segments of similar orientation. These limitations are a function of the structure of the image, and the size of the initial $\nabla^2 G$ operator used to derive the zero-crossings.

The first limitation is a consequence of the confounding of orientation and velocity of the zero-crossings. The detection of a discontinuity, such as that of Figure 5.1, requires that there exist zero-crossing segments whose orientation either lies between the directions of velocity of the two adjacent surfaces (yielding a change in the sign of $v^{\perp}(s)$), or is such that there is a significant change in the magnitude of $v^{\perp}(s)$. If the image structure is such that there exists relatively few zero-crossing segments at the necessary orientations, then detection of the boundary is hindered. In natural images, the local distribution of zero-crossing orientations is often nonuniform, so that the early detection of a discontinuity may not always be possible. Even with a roughly uniform distribution, as the change in direction and magnitude of velocity across the boundary becomes smaller, the density of zero-crossing segments at the necessary orientations may decrease. Thus, the ability to detect a discontinuity from the initial measurements of $v^{\perp}(s)$ along zero-crossing contours, is dependent on the image structure, and generally decreases with smaller changes in velocity across the boundary.

A second limitation of this approach arises from the fact that the density of the zero-crossings is constrained by the size of the initial $\nabla^2 G$ operator. For the case of larger operators, zero-crossing contours are more sparse, so that boundaries may not be localized as well, and very small objects against a textured background may go undetected. For the smaller operators, measurements are more dense, allowing more accurate localization of a boundary, and the discrimination of smaller objects.

As we mentioned in Section 4.4.2, however, for the case of discrete presentation, the smaller operators can only yield reliable measurements of movement for small displacements in the image. There is a resulting trade-off between the range of displacements allowed, and the resolution with which discontinuities can be analyzed. In addition to the general dependence of the density of zero-crossings on operator size, there may also be a nonuniform density of zero-crossings across the image, due to variations in image structure.

### 5.4. Translation versus Rotation

Suppose that two adjacent surfaces $\mathbf{S}_1$ and $\mathbf{S}_2$, undergo pure translation with velocities $\mathbf{V}_1$ and $\mathbf{V}_2$, respectively. Let $P_1$ and $P_2$ denote two adjacent image patches, to which these surfaces project. In this case, if only the sign of $v^{\perp}(s)$ is examined, for zero-crossing segments whose orientation lies between the directions of $\mathbf{V}_1$ and $\mathbf{V}_2$, there will be one uniform sign in $P_1$, and the opposite uniform sign in $P_2$. This is illustrated in Figure 5.2. In Figure 5.2a, the central square patch of the random dot pattern moves in a direction opposite the direction of motion of the surrounding area, as the arrows illustrate. (The dotted black lines only indicate the location of the boundary, and are invisible in the actual images.) In this case, zero-crossing segments of all orientations, except those that are horizontal, indicated by the grey area of Figure 5.2c, yield a change in the sign of $v^{\perp}(s)$ across the boundary. The information provided by one representative orientation is shown in Figure 5.2f. In Figure 5.2b, the directions of motion of the central patch and its surround differ by only 90°; in this case, only those zero-crossing segments whose orientation lies in the grey area of Figure 5.2d yield a change in the sign of $v^{\perp}(s)$ across the boundary, as shown by the sample orientation of Figure 5.2g. Zero-crossing segments whose orientation lies in the range indicated in Figure 5.2e yield a single sign of $v^{\perp}(s)$ across the entire image, as shown in Figure 5.2h. The magnitude of $v^{\perp}(s)$ can be considered as well. For zero-crossing segments of a particular orientation, either this magnitude will be constant across the entire image, or there will be one uniform magnitude over the central square, and a different uniform magnitude over the surrounding area.

The situation is quite different when surfaces undergo rotation. In the random dot pattern of Figure 5.3a, a central circular patch of dots is rotated in the clockwise direction, and the background is translated (as in the previous example, the border is

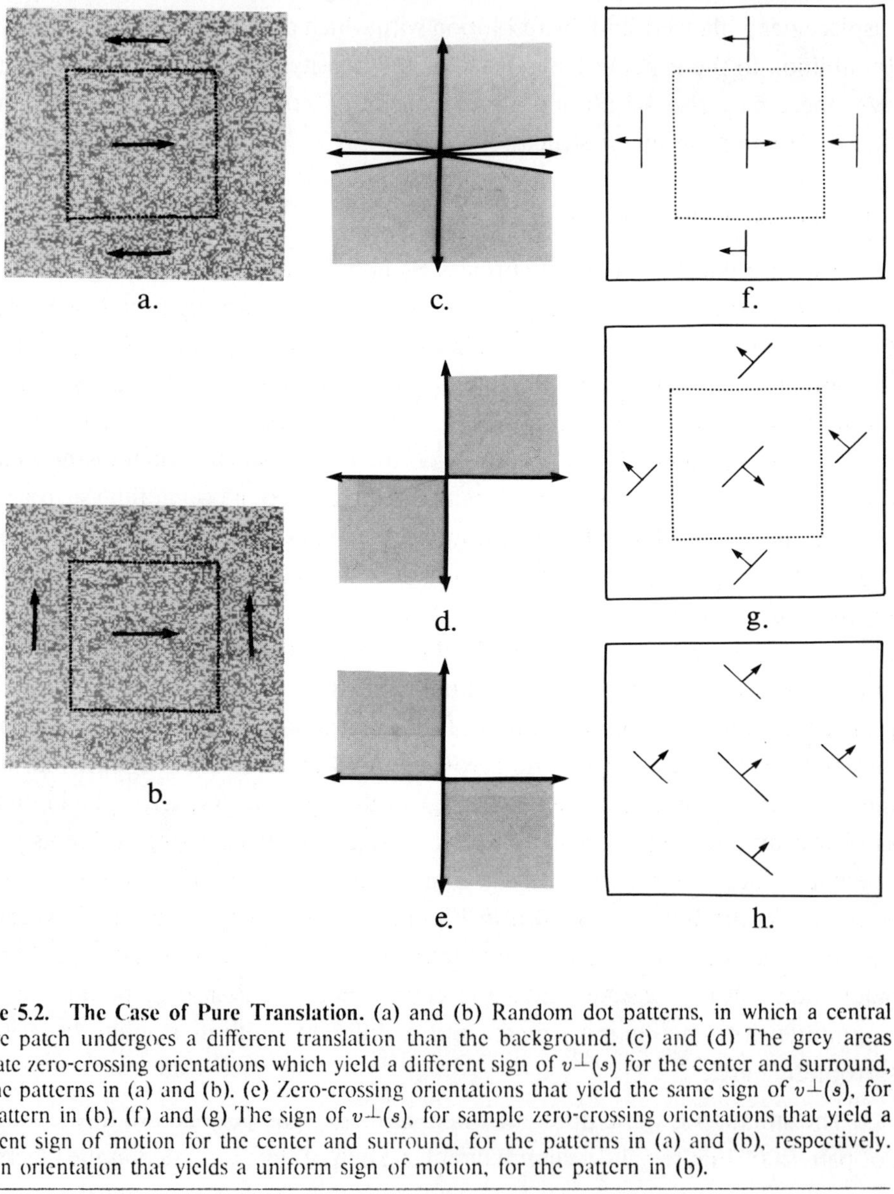


**Figure 5.2. The Case of Pure Translation.** (a) and (b) Random dot patterns, in which a central square patch undergoes a different translation than the background. (c) and (d) The grey areas indicate zero-crossing orientations which yield a different sign of $v^{\perp}(s)$ for the center and surround, for the patterns in (a) and (b). (e) Zero-crossing orientations that yield the same sign of $v^{\perp}(s)$, for the pattern in (b). (f) and (g) The sign of $v^{\perp}(s)$, for sample zero-crossing orientations that yield a different sign of motion for the center and surround, for the patterns in (a) and (b), respectively. (h) An orientation that yields a uniform sign of motion, for the pattern in (b).

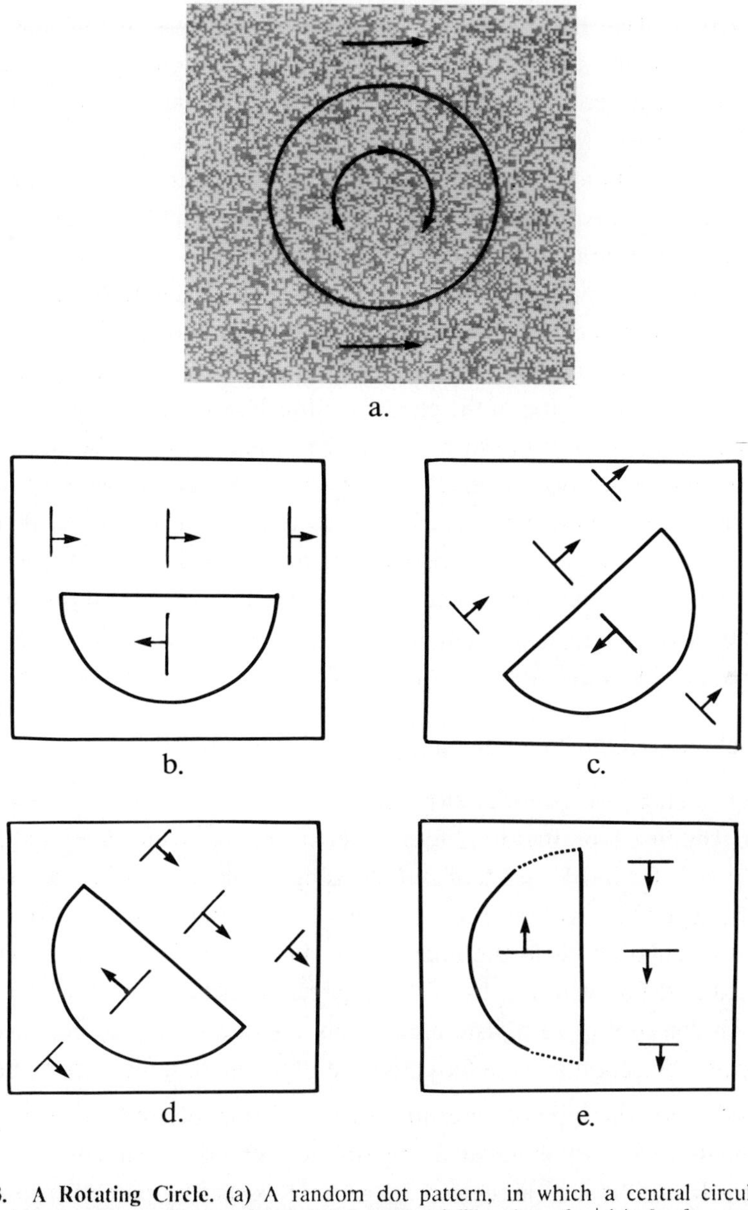


**Figure 5.3. A Rotating Circle.** (a) A random dot pattern, in which a central circular area is rotated, and the background is translated. (b) through (e) The sign of $v^{\perp}(s)$, for four zero-crossing orientations. The solid lines indicate the border between the two signs of $v^{\perp}(s)$.

invisible in the static frames). The human visual system easily detects this boundary, even in a brief flash of two frames. In this case, changes in the sign of $v^{\perp}(s)$ occur in areas of the image where there is no real discontinuity in the velocity field, and different zero-crossing orientations yield a change in the sign of $v^{\perp}(s)$ along different parts of the actual boundary. The signs of $v^{\perp}(s)$, provided by some representative orientations, are shown in Figures 5.3b through e. Vertically oriented zero-crossing segments, for example (see Figure 5.3b), yield a change in the sign of $v^{\perp}(s)$ along the bottom half of the circle, and there is an additional change in sign along the horizontal diameter of the circle. The magnitude of $v^{\perp}(s)$ varies smoothly around the central circular patch, and is zero along this diameter. Other zero-crossing orientations yield measurements of $v^{\perp}(s)$ that reveal different parts of the rotating circle. At the top of the circle, the direction of velocity inside the central circle is identical to that of the surround; if the magnitudes are also the same, then no real discontinuity exists here. Thus, the smoothly varying velocity fields that result from surfaces undergoing rotation pose two problems for the detection of discontinuities from the initial motion measurements. First, if only the sign of $v^{\perp}(s)$ is considered, then changes in sign can occur where there is no real discontinuity in the velocity field; second, measurements from zero-crossings in different orientation ranges may provide a different description of the surface boundaries.

### 5.5. Possible Algorithms for Detecting Motion Discontinuities

In this section, we consider two general types of algorithm for discontinuity detection. The first type performs local comparisons between individual measurements of $v^{\perp}(s)$, for nearby pairs of zero-crossing segments of the same orientation. The second type of algorithm utilizes changes in local statistics of $v^{\perp}(s)$, computed over small neighborhoods in the image. One simple algorithm of the second type is presented in detail, with results of the application of the algorithm to a complex dynamic random dot pattern. The aim of this exercise is only to demonstrate the feasibility of a detection scheme based on the initial motion measurements.

Consider first the type of algorithm that performs local comparisons between measurements of $v^{\perp}(s)$ obtained at nearby zero-crossing segments of the same orientation. If a difference in sign or magnitude exists, the pair of measurements may signal a local motion discontinuity. In order to detect boundaries that occur at any orientation in the image, it is necessary to compare one measurement of

$v^{\perp}(s)$ with other measurements of $v^{\perp}(s)$, obtained anywhere within a circular neighborhood of the first measurement. In a biological system, one can imagine such comparisons performed by a local inhibitory/excitatory network. A movement detector that signals a zero-crossing segment with a particular orientation and sign (or magnitude) of $v^{\perp}(s)$ could inhibit other detectors of the same type, within some neighborhood, and excite detectors that signal a zero-crossing segment with the same orientation but opposite sign (or different magnitude) of $v^{\perp}(s)$. The activity of such a network should, in principle, result in the inhibition of measurements within a patch of uniform motion, and enhancement of measurements along motion discontinuities. This type of algorithm is similar to that proposed by Reichardt and Poggio (1979) for relative movement detection by the visual system of the housefly.

The second approach to discontinuity detection described above utilizes changes in local statistics of $v^{\perp}(s)$, computed over small neighborhoods in the image. We present one algorithm, based on the detection of changes in the average sign of $v^{\perp}(s)$. The algorithm can be extended to incorporate the magnitude of $v^{\perp}(s)$ as well; however, magnitude measurements are generally less reliable than measurements of the sign of $v^{\perp}(s)$, particularly when derived from natural images. We therefore chose first to consider an algorithm that utilizes only the sign of $v^{\perp}(s)$.

In order to detect discontinuities that arise from small changes in the direction of velocity, it is necessary to compute the average sign of $v^{\perp}(s)$ for measurements obtained from zero-crossing segments within small ranges of orientation. (It appears, for example, that the human visual system can detect a boundary, across which the direction of velocity changes by only $25°$; this result is discussed in Chapter 7.) Measurements obtained within one orientation range, however, cannot provide good localization of the boundary, due to the sparseness of the available zero-crossings. This raises the problem of combining boundary evidence across different orientation ranges, in order to provide more reliable detection and localization of the boundaries. In the particular algorithm to be described here, motion discontinuities are first detected independently, from measurements of $v^{\perp}(s)$ obtained for a set of disjoint zero-crossing orientation ranges. If information from two or more orientation ranges signals the same discontinuity, then this information is combined to provide better localization of the boundary.

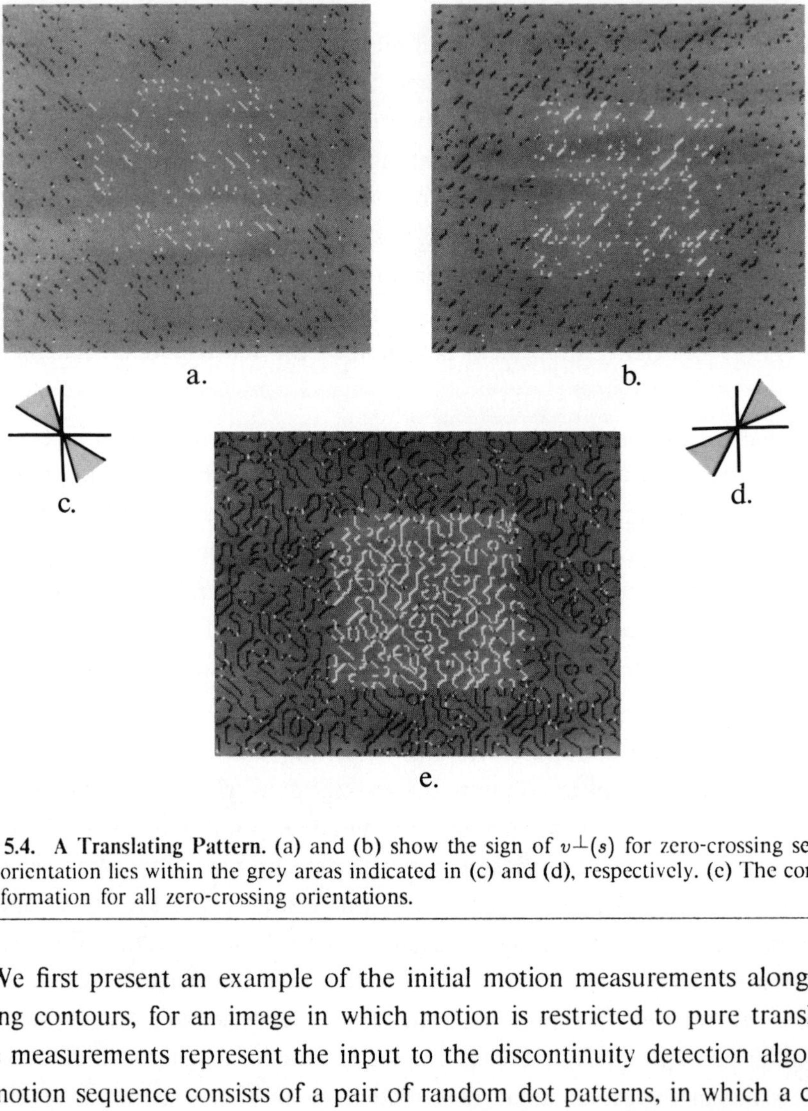


**Figure 5.4. A Translating Pattern.** (a) and (b) show the sign of $v^{\perp}(s)$ for zero-crossing segments whose orientation lies within the grey areas indicated in (c) and (d), respectively. (e) The combined sign information for all zero-crossing orientations.

We first present an example of the initial motion measurements along zero-crossing contours, for an image in which motion is restricted to pure translation. These measurements represent the input to the discontinuity detection algorithm. The motion sequence consists of a pair of random dot patterns, in which a central square patch is shifted to the right in the second pattern, while the background is shifted to the left. The two frames were convolved with a $\nabla^2 G$ operator, the temporal derivative, $\frac{\partial}{\partial t}(\nabla^2 G * I)$, was computed by subtracting the two filtered patterns, and $v^{\perp}(s)$ was computed using the method described in Section 4.4.2. Figures 5.4a and

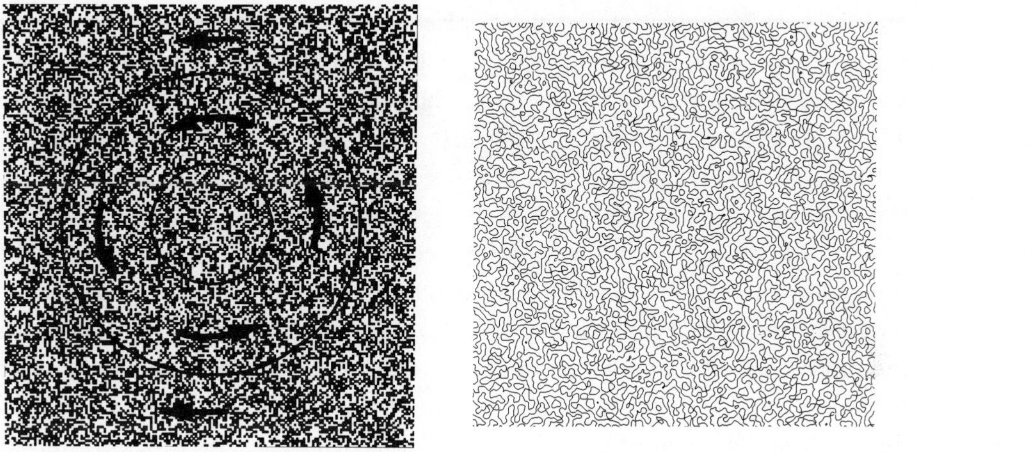

a. b.

**Figure 5.5. Dynamic Random Dot Pattern.** (a) An annulus is rotated counterclockwise, while the background is translated to the right. The boundaries are not visible in the individual frames. (b) The zero-crossing contours, derived from the convolution of the first frame with a $\nabla^2 G$ operator.

b display the sign of $v^{\perp}(s)$ for two orientation ranges. In Figure 5.4a, zero-crossing segments whose orientation lies within the grey area of Figure 5.4c, and whose sign of $v^{\perp}(s)$ corresponds to motion directed obliquely upward and to the right, are displayed in white. Those of opposite sign are displayed in black. In Figure 5.4b, zero-crossing segments whose orientation lies within the grey area of Figure 5.4d, and whose sign of $v^{\perp}(s)$ corresponds to motion directed obliquely down and to the right are displayed as white, while those of the opposite sign are black. The combined sign information from the full range of zero-crossing orientations (excluding only those that are roughly horizontal), is shown in Figure 5.4e. While there are a few scattered incorrect measurements, the sign information alone clearly distinguishes the central square from the surrounding area.

In the previous section, it was noted that the detection of discontinuities is more difficult when surfaces are allowed to rotate. Therefore, to illustrate the basic steps of the algorithm, and test its performance, a dynamic random dot pattern, in which a central annulus rotates against a translating background, will be used. The first image, shown in Figure 5.5a, contains $768 \times 768$ picture elements (pixels), with a dot size of $4 \times 4$ pixels. The radii of the inner and outer circles are 100 and

260 pixels, respectively. The annulus is rotated $1°$ in the counterclockwise direction, and the background is translated two pixels to the left. The largest displacements in the image, along the outer border of the annulus, are 5 pixels. The patterns were convolved with a $\nabla^2 G$ operator whose central diameter, $w$, was 12 pixels. The resulting zero-crossings for the first frame are shown in Figure 5.5b. The zero-crossings were divided into four disjoint orientation ranges, each $45°$ wide (local orientation was computed from the gradient of the filtered image). The choice of orientation ranges is flexible, and they need not be disjoint. The signs of $v^{\perp}(s)$ for the four orientation ranges are displayed in Figures 5.6a through d. In each case, the particular orientation range is shown below the zero-crossing representation, with the direction of the positive and negative sign of $v^{\perp}(s)$ indicated. Zero-crossings with a negative sign of $v^{\perp}(s)$ are displayed in black, those with a positive sign are displayed in white, and the background is shown in grey. As expected, the four orientation ranges reveal different parts of the rotating annulus.

Within the four orientation ranges, the local average sign of $v^{\perp}(s)$ was then computed, using a neighborhood size of $41 \times 41$ pixels, centered on each point in the image. This neighborhood size is slightly smaller than the overall extent of the initial $\nabla^2 G$ operator. To compute this average sign, zero-crossings with a positive sign of $v^{\perp}(s)$ were assigned a value of 200, and those with a negative sign were assigned the value 100. The resulting average sign then ranges in value from 100 (black) to 200 (white). The average signs for the measurements shown in Figures 5.6a through d are shown in Figures 5.7a through d, respectively. As expected, the average sign for the four orientation ranges reveals different parts of the rotating annulus. Along the inner and outer edges of the annulus, the average sign of $v^{\perp}(s)$ changes sharply. Within each orientation range, the sign of $v^{\perp}(s)$ also changes across a diameter of the annulus; however, the rate of change in average sign is slow here, and the magnitude of $v^{\perp}(s)$ is small.

Suppose there is a step change in the sign of $v^{\perp}(s)$, within some orientation range. Such a change is shown in one dimension in Figure 5.8a. Let the neighborhood size, used for computing average sign, be denoted by $d$. If the local distribution of zero-crossing measurements is roughly uniform, the average sign of $v^{\perp}(s)$ changes from negative to positive over the distance $d$, as shown in Figure 5.8b. Let the values associated with positive and negative signs of $v^{\perp}(s)$ be denoted by $p$ and $n$, respectively. In this ideal case, the average sign of $v^{\perp}(s)$ at the location of the

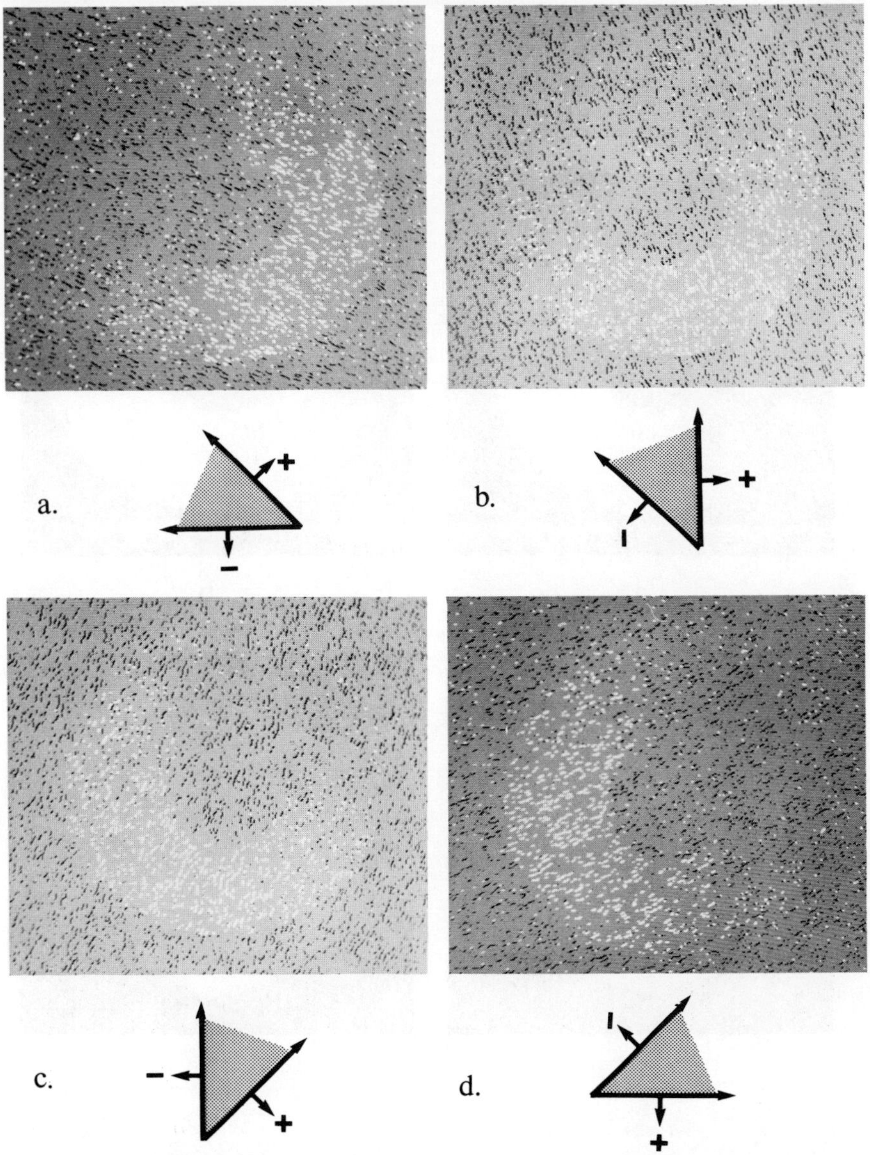


**Figure 5.6. The Signs of $v^{\perp}(s)$.** (a) through (d) show the sign of $v^{\perp}(s)$ for zero-crossing segments whose orientation lies within the ranges shown. Zero-crossings yielding a positive sign of $v^{\perp}(s)$ are displayed as white, and those with negative sign are black.

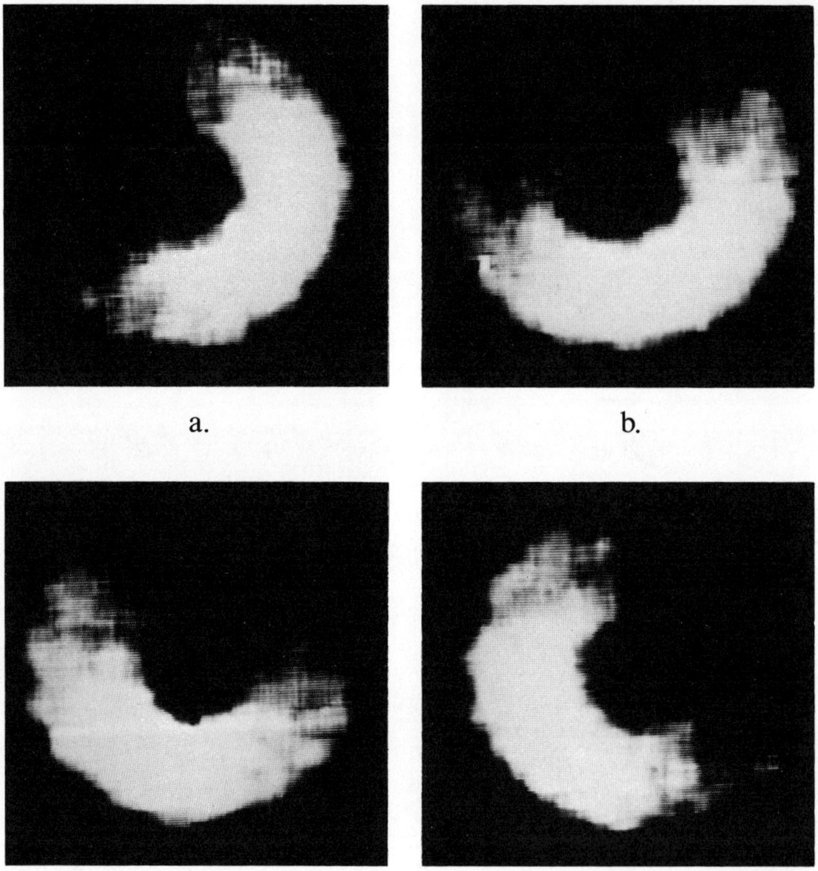

a.  b.

c.  d.

**Figure 5.7. Average Sign of $v^{\perp}(s)$.** (a) through (d) show the average sign of $v^{\perp}(s)$ for zero-crossing segments whose orientation lies within the ranges indicated in Figure 5.6. White represents an average value of 200 (positive sign), and black represents a value of 100 (negative sign).

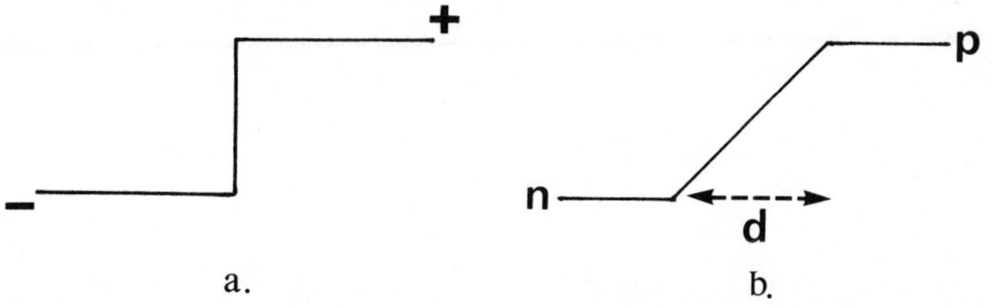


Figure 5.8. An Ideal Discontinuity. (a) A one-dimensional step change in the sign of direction of motion. (b) The change in average sign of direction across the edge, given a neighborhood size *d*.

discontinuity is equal to the mean value, $\frac{p+n}{2}$. For the descriptions of average sign, shown in Figure 5.7, the mean value is 150; an average sign equal to this mean value may signal a local change in the sign of $v^{\perp}(s)$, for the particular range of zero-crossing orientations. As we noted above, average sign changes sharply at the inner and outer borders of the annulus. The sign of $v^{\perp}(s)$ also changes along diameters of the annulus, but the change is slow, and the average sign of $v^{\perp}(s)$ fluctuates about the mean value over an extended area. The local variance in the sign of $v^{\perp}(s)$ may then be used to distinguish the two types of change.

In Figures 5.9a through d, we plot points in the average sign descriptions of Figures 5.7a through d, for which the following is true: (1) the average sign of $v^{\perp}(s)$ crosses the mean value, 150, and (2) local variance in average sign is large (determined by the expected variance for a step change in sign of direction of motion). The contours roughly isolate different parts of the inner and outer borders of the annulus. In the final step of the algorithm, local information about average sign is then averaged across orientation ranges, when two or more ranges yield a discontinuity in roughly the same location in the image. In Figure 5.10, points at which this combined description of average sign crosses the mean value are displayed. Localization of the inner and outer borders of the annulus is much improved. Discontinuities are not detected at the top of the annulus, where the direction of velocity of the annulus is the same as that of the surround. A few spurious contours at the center of the annulus also appear in the result.

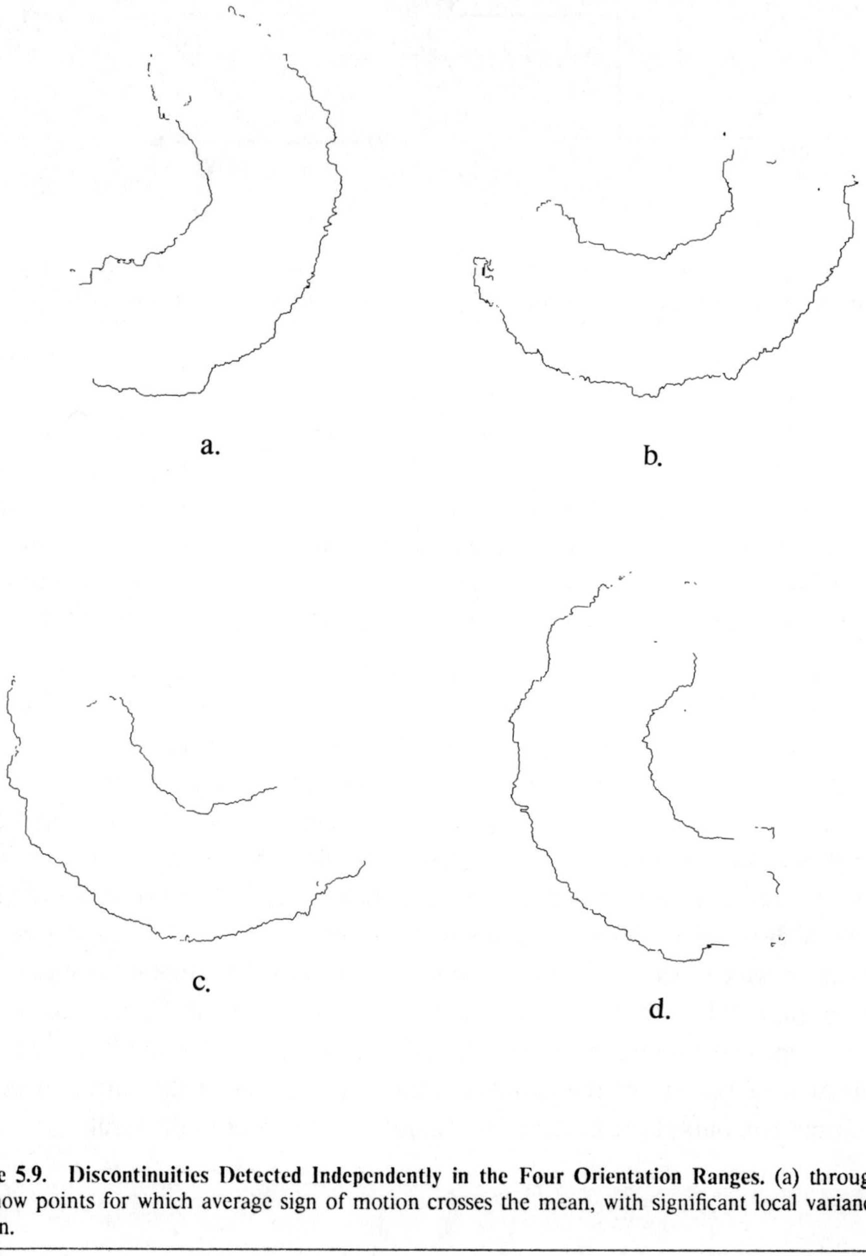

Figure 5.9. Discontinuities Detected Independently in the Four Orientation Ranges. (a) through (d) show points for which average sign of motion crosses the mean, with significant local variance in sign.

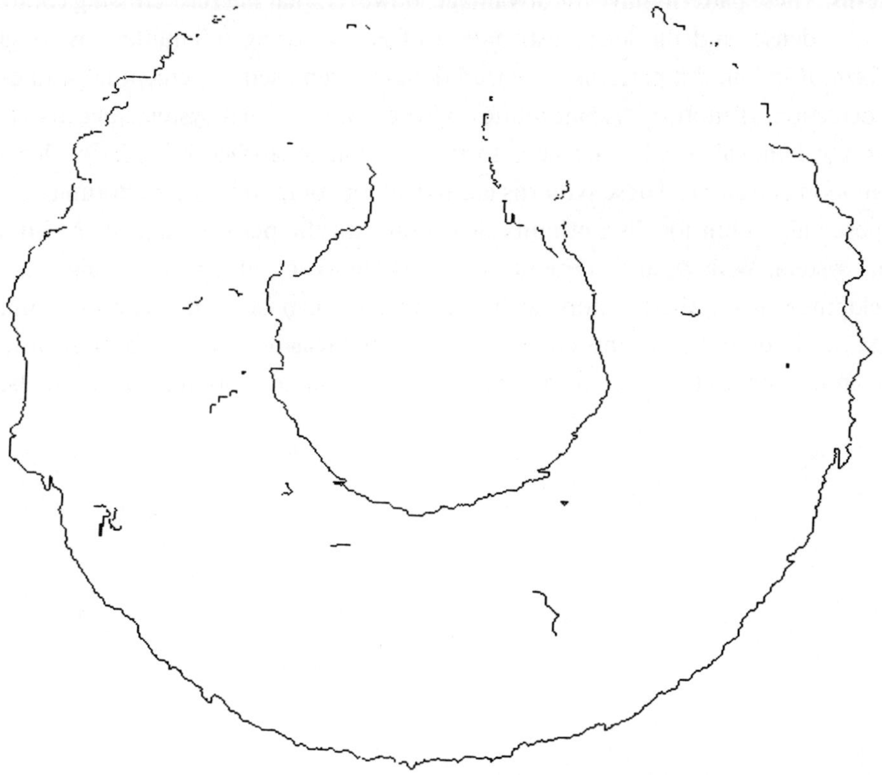

**Figure 5.10. The Final Discontinuities.** The results of combining boundary information across the four orientation ranges.

The basic steps of the algorithm can be summarized as follows:

(1) The average sign of $v^{\perp}(s)$ is computed, within a set of zero-crossing orientation ranges.

(2) Discontinuities are detected within separate orientation ranges, by locating points at which the average sign of $v^{\perp}(s)$ crosses the mean value, with significant local variance in sign.

(3) If the same discontinuity is detected by two or more orientation ranges, the average sign is combined across orientations, and the mean of the result is used to locate the discontinuities.

The algorithm is fairly simple, and performs reasonable well on dense random dot patterns. These patterns have the advantage, however, that the zero-crossing contours are very dense, and the local distribution of zero-crossing orientations is roughly uniform. Random dot patterns have traditionally been used in perceptual studies of the detection of motion discontinuities by the human visual system, because they allow the removal of all other cues to the location of surface boundaries, leaving the motion cue alone. These patterns are useful for comparing the performance of a proposed algorithm for discontinuity detection with the performance of the human visual system. With regard to the present work, however, much further algorithmic development is required, before such a comparison can be made. The preliminary investigation described in this chapter supports the feasibility of performing at least a rough discontinuity detection on the basis of the initial motion measurements.

# 6. PERCEPTUAL STUDIES OF MOTION MEASUREMENT

In the remaining chapters of this book, we examine a number of questions regarding the measurement of motion in the human visual system, that arise from the computational study presented in the previous chapters. We begin by discussing some perceptual studies relevant to the velocity field computation. The idea that the measurement of motion requires explicit computation, beyond the initial measurements obtained from the changing image, can be motivated by simple examples. Figure 6.1 illustrates a circle and square undergoing pure translation. Local motion measurements from the image can provide only the component of motion in the direction perpendicular to the local orientation of the contours, as shown in Figures 6.1c and d. The perception of movement of the circle or square corresponds to pure translation, indicated by the set of velocity vectors in Figures 6.1a and b. Other examples are given by the perception of movement of the smooth curves presented in Section 4.4.1. While it is not yet clear whether the human visual system explicitly represents the local velocity field along the contours, there are properties of their perceived motion that are not explicit in the initial motion measurements. Examples include the radial movement of the rotating logarithmic spiral, and the downward movement of the circular helix (analogous to the perceived movement of the stripes on a barberpole). All of these phenomena indicate that motion measurement requires some type of computation beyond the initial detection of movement in the changing image.[1]

There are two primary questions that we address through perceptual studies. The first concerns the input representation for the velocity field computation: *Could the initial detection of motion take place at the location of zero-crossings in the output of the* $\nabla^2 G$ *convolution, and is perceived movement consistent with at least the sign of* $v^{\perp}(s)$ *along these zero-crossing contours?* The second question concerns the additional assumptions that allow the human visual system to compute motion uniquely: *Is the perception of motion consistent with the computation of a velocity field of least variation?* These two questions are addressed in Sections 6.1 and 6.2, respectively. The remaining sections briefly describe four additional aspects of motion measurement for psychophysical investigation: (1) the derivation

[1]Movshon *et al.* (1983) provide direct psychophysical evidence for the existence of a two-stage process for motion measurement in the human visual system.

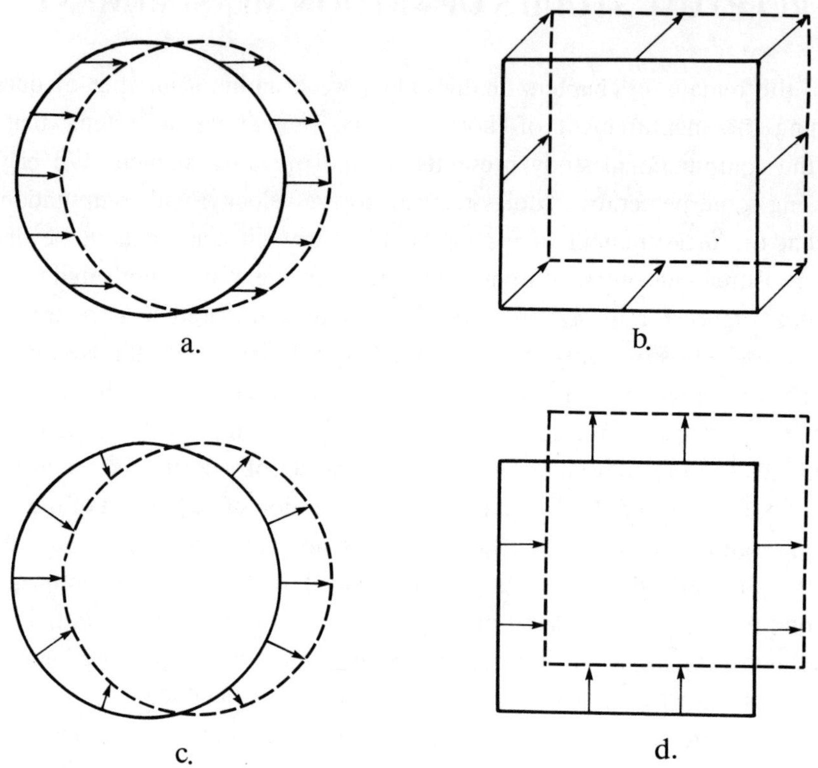


**Figure 6.1. Two Translation Examples.** (a) and (b) The true velocity fields for a translating circle and square. (c) and (d) The initial perpendicular components of velocity.

of further motion constraints from eye movements, (2) the tracking of localizable features in the image, (3) the role of short range and long range motion processes, (4) the use of multiple channels with different spatial and temporal frequency characteristics, and (5) the independence of the measurement of motion, and derivation of three-dimensional structure from motion.

## 6.1. The Measurement of Motion Along Zero-crossing Contours

In Sections 6.1.1 and 6.1.2, we examine a number of short range motion phenomena, and ask whether the perception of motion is consistent with the movement of zero-crossings contours. This analysis utilizes a very simple model of early visual processing. The intensity function corresponding to the visual stimulus is first convolved with an ideal $\nabla^2 G$ operator (or its one-dimensional counterpart, $D^2 G$). It is assumed that this initial filtering also embodies a temporal smoothing function, which is limited in time. In the output of this filtering stage, the instantaneous movement of the zero-crossings is observed, and compared with the perception of motion of the stimulus. The actual early processing stages in human vision are certainly more complex, but this simple model captures some of the proposed aspects of that processing, which are sufficient to account for many of the short range phenomena. This analysis does not attempt to provide proof that the initial measurement of motion takes place at the location of zero-crossings. There are alternate methods for motion measurement that are also consistent with the phenomena discussed here. In addition, there may exist multiple pathways in the human visual system with motion measurement capabilities, which may use different mechanisms for performing this measurement. In Section 6.1.3, it is argued that further discrimination between alternate methods requires a commitment to more detailed models of the early visual processing stages, and the motion detection mechanisms.

### 6.1.1. The Short Range Motion Phenomena

In this section, we examine Anstis' reversed phi phenomena (Anstis 1970, 1980; Anstis and Rogers 1975), the dynamic random dot patterns studied by Julesz (1971) and Braddick (1973, 1974, 1980), a dynamic visual noise demonstration described by Morgan and Ward (1980), experiments based on the Ternus configuration (Ternus 1926; Pantle and Picciano 1976; Petersik and Pantle 1979), and an additional demonstration presented by Ramachandran and Anstis (1983; Anstis 1983). The main interest here is in the consistency of the perception of motion with the movement of zero-crossing contours. More generally, these demonstrations suggest that short range motion measurements provide powerful constraint for the computation of motion, which cannot easily be overridden, if at all, by other factors, such as a long range motion process, or cognitive processes (a point also made by Mather and Anstis (1983)).

The limitations on the spatial and temporal displacements that characterize the short range motion process may be a function of the limited spatial and temporal extent of the initial filtering of the image. The size of the $\nabla^2 G$ operator is represented by $w$, the diameter of its central positive region; the size of its one-dimensional counterpart, $D^2G$, is given by $w_x$, the width of its central positive region. For the analysis shown here, no particular model is assumed for the temporal properties of the initial operators (although quantitative models have been proposed, and are discussed briefly in Section 6.1.3).

Consider first Anstis' reversed phi demonstration, illustrated in Figure 6.2. The basic stimuli are shown in Figures 6.2a and b. The first frame consists of an edge or dot of one contrast, and the second an edge or dot of opposite contrast, slightly displaced. The two frames are either presented in rapid succession, or there is a fading from one to the other. Figures 6.2c and d show samples of the sequence of one-dimensional profiles that would appear during a fade from frame 1 to frame 2 in the dot and edge demonstrations. If the physical displacement of the two edges or two dots is less than about 10′ of visual arc, and the time interval between the two frames is less than about 50 milliseconds (msec), motion is perceived in the direction opposite the physical displacement, as indicated by the arrow between the two frames in Figures 6.2a and b. With larger displacements, the change is not perceived as movement.

In Figure 6.3a, the sequence of intensity profiles shown in Figure 6.2d is redisplayed, with each profile represented by a different pattern of lines and dots. Figure 6.3c shows the convolutions of the profiles of Figure 6.3a with the one-dimensional $D^2G$ operator (a similar analysis appears in Anstis and Rogers (1975)). In this example, if we let the edge displacement be denoted by $d$, then the size of the $D^2G$ operator is $w_x = 2d$. The five convolution profiles are shown superimposed. As we move from profile 1 to profile 5, the zero-crossings of the corresponding convolutions shift to the left, as indicated by the arrows. The physical displacement, however, is to the right. Thus, the perception of motion to the left is consistent with the movement of zero-crossings (this was also noted by Marr and Ullman (1981)). Anstis and Rogers (1975) found the perception of the speed of movement in the opposite direction to be inversely proportional to the size of the displacement. In Figure 6.3b, a similar sequence of intensity profiles is shown, with a larger physical displacement between the two edges. In the corresponding convolution profiles of Figure 6.3d, the extent of motion of the zero-crossings is less, again consistent with the perception. In the computer simulations, significant movement of the zero-crossings was obtained for edge displacements as large as about $w_x$. When $d$ was increased beyond $w_x$, the motion of zero-crossings became

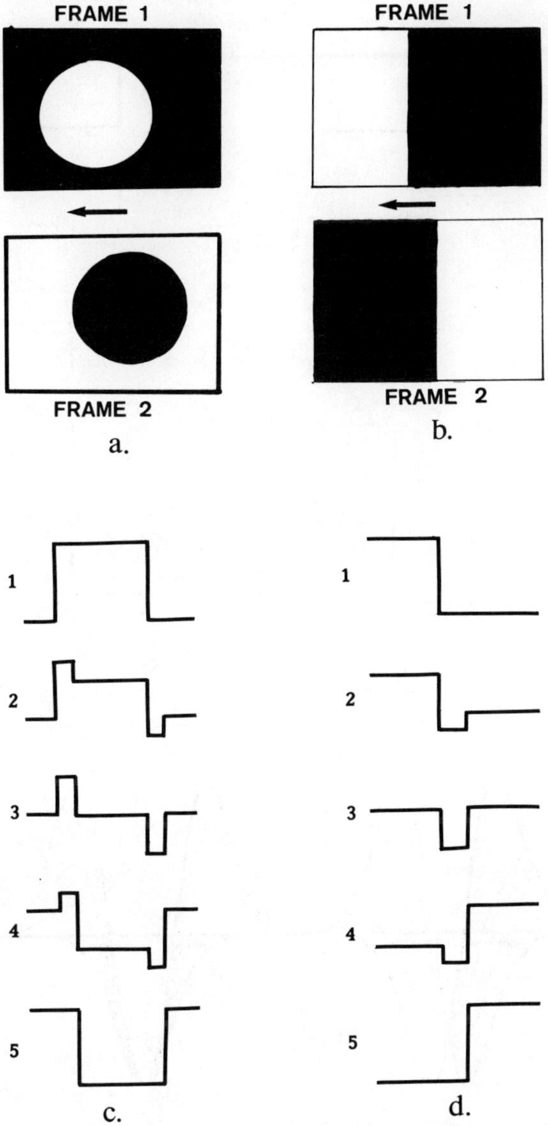


**Figure 6.2. The Reversed Phi Demonstration.** (a) and (b) illustrate the two frames for the dot and edge demonstrations. The arrows indicate the perceived direction of movement. (c) and (d) illustrate the one-dimensional profiles, at five different instants of time, when there is a fade between the two frames.

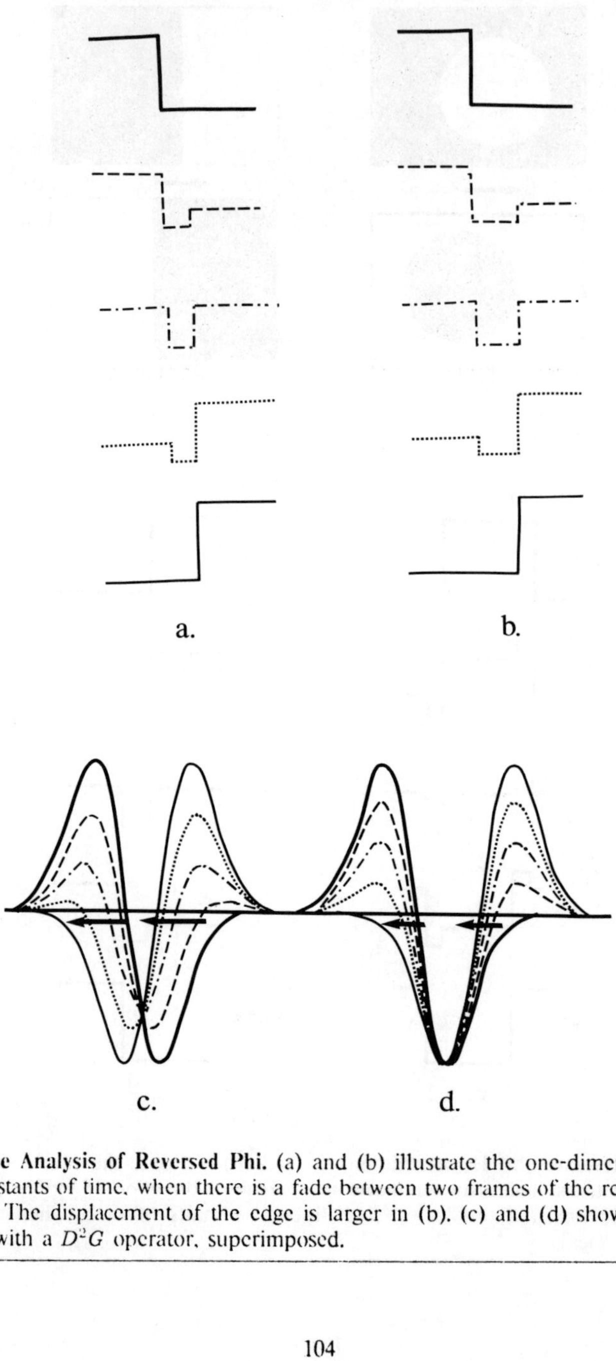


**Figure 6.3. The Analysis of Reversed Phi.** (a) and (b) illustrate the one-dimensional profiles, at five different instants of time, when there is a fade between two frames of the reversed phi pattern of Figure 6.2b. The displacement of the edge is larger in (b). (c) and (d) show the convolutions of the profiles with a $D^2G$ operator, superimposed.

negligible (for example, for an edge displacement of $d = \frac{3w_x}{2}$, the total displacement of the zero-crossings was roughly $.05w_x$). The introduction of temporal smoothing of the visual input influences the rate of movement of the zero-crossings, but does not change their qualitative behavior.

For the case of moving random dot patterns, Braddick (1973, 1974, 1980; Baker and Braddick 1982a,b) demonstrated that coherent motion of the pattern is perceived as long as the displacement of the dots, and time interval between frames, is within the short range limits. Outside this range, incoherent motion is observed, with patches of the pattern moving in different directions. In this case, the analysis of zero-crossings was begun with one-dimensional slices of a moving random dot pattern. The profiles were convolved with the one-dimensional $D^2G$ operator, and the direction of motion of the resulting zero-crossings was computed by comparing the signs of the spatial and temporal derivatives of the filtered pattern (it is not critical that the direction of movement be computed in this manner). As long as the displacement of the pattern was less than about $w_x$, this computation predicted coherent motion of the zero-crossings in the correct direction. As the displacement reached roughly $\frac{3w_x}{2}$, the motion of the zero-crossings became incoherent; that is, it occurred in both directions over the entire pattern. The behavior of the zero-crossings in this analysis is qualitatively consistent with the spatial aspects of the perception of motion of the dynamic random dot pattern. Within a limited range of displacements (up to $10 - 15'$ of visual arc), coherent motion of the pattern is perceived. Beyond this range, the pattern no longer appears to move coherently, breaking up into small patches moving in different directions.

An additional short range demonstration, which we did not analyze in detail, was presented by Morgan and Ward (1980). Dynamic visual noise was created on an oscilloscope screen by transiently displaying small dots at a rapid rate. Half of the dots were placed at random, and the other half were paired with the first, with each displayed at a particular spatial and temporal displacement from its partner. The spatial and temporal displacements were constant for all pairs of dots. Morgan and Ward found that as long as the spatial displacements were less than about $18'$ of visual arc, and temporal displacements were less than $60 - 100$ msec (this varied with the contrast of the display), then the field of dots appeared to move coherently in a single direction. Otherwise, the field appeared as dynamic noise. The behavior of the zero-crossings for this visual stimulus was not examined in detail; however, we predict that this behavior is consistent with the perception of movement in these displays. This prediction is based on the analysis of a similar demonstration described by Anstis and Ramachandran (Anstis 1983), discussed in Section 6.2.1 and illustrated in Figure 6.14. When the spatial and temporal displacements are within some limit, determined by the size of the $\nabla^2 G$ operator and its temporal properties, coherent movement of the zero-crossing contours in the direction of actual displacement between pairs of dots is expected. If for some dots, a nonpartner also appears within a sufficiently small spatial and temporal interval, the local movement of the zero-crossing contours would be corrupted, yielding a local velocity direction that differs from the coherent direction of movement. The judgements of movement direction in this experiment were presumably based on an overall impression of dot flow, so may not have been degraded by local deviations in the direction of movement caused by the accidental proximity of nonpaired dots. When the spatial and temporal displacements are outside the range of a particular $\nabla^2 G$ operator, two possible behaviors of the local zero-crossings are expected to occur. If no dots appear within some limited spatial and temporal range of a particular dot, the local zero-crossing contour would remain stationary and eventually disappear. Otherwise, the contour would move in a direction determined by the chance appearance of nonpartner dots. In either case, no coherent movement of the zero-crossing contours in one direction is expected. We therefore predict that the behavior of the zero-crossing contours would be consistent with the perception of movement in the dynamic noise displays used by Morgan and Ward.

It is difficult to make quantitative comparisons between the results of the computer simulations of the behavior of zero-crossings for the short range phenomena, and the psychophysically measured limitations of the short range process. This is in part because the quantitative aspects of the initial operators in the human visual system are not fully known. With regard to their spatial characteristics, Wilson and Bergen (1979; Wilson and Giese 1977) have proposed one quantitative model, based on studies of threshold spatial vision, that suggests that at each location in the visual field, the retinal image is processed through roughly four channels. The proposed shape of the underlying operators is the difference of two Gaussians; in Wilson and Bergen's model, the width of the central positive region of the one-dimensional projections of these operators, $w_x$, is roughly $3'$, $6'$, $11'$ and $21'$ of visual arc in the central fovea. The size of the operators increases linearly with eccentricity.[2] In the short range demonstrations, the areas of the visual patterns that were critical for the particular task were generally limited in size to the foveal region (for example, the central patches were 2.25° × 2.25° in Morgan and Ward's (1980) experiments, and 2° × 4° in Braddick's (1974) studies). In the computer simulations of the reversed phi phenomena, it was found that a $D^2G$ operator of size $w_x$ yields significant motion of the zero-crossings in the reverse direction for edge displacements up to about $w_x$. For the random dot patterns, coherent motion of the zero-crossings occurred for displacements up to about $w_x$. The spatial limits that are demonstrated empirically for the short range phenomena are within a range that is expected for the 6′ or 11′ channels in Wilson and Bergen's model. This is only a rough comparison, however, because the psychophysically measured short range limits are not sharply defined, and it is not yet possible to model the precise spatial and temporal characteristics of the early operators in the human visual system. With regard to their temporal properties, the smaller (high spatial frequency sensitive) channels exhibit a sustained response to temporal modulations in the intensity function, while the larger (low spatial frequency sensitive) channels exhibit transient responses (Tolhurst 1973, 1975; Kulikowsi and Tolhurst 1973; Breitmeyer and Ganz 1977; Watson and Nachmias 1977). This may suggest that the channels play different roles in the measurement of motion. The spatiotemporal properties of the channels, and their role in motion measurement, are discussed further in Sections 6.1.3 and 6.6.

Ramachandran and Anstis (1983; Anstis 1983) present a demonstration that illustrates the powerful constraint imposed by short range motion measurements. They began with an ambiguous motion display consisting of four dots, shown in Figure 6.4a. The filled circles form the first frame, and the open circles form the second. With a large distance between the dots (1° of visual arc), there is equal probability of seeing the dots as moving vertically (Figure 6.4b) or horizontally (Figure 6.4c). The pattern of four dots was then embedded in a horizontal sequence of flashing dots that mimicked two spots moving in opposite directions along parallel horizontal paths. This new motion sequence is illustrated in Figure 6.4d. The numbers indicate the order of positions at which the dots appeared (the original

[2] More recent studies have revealed at least one smaller operator in the central fovea (Marr, Poggio and Hildreth 1980; Watson 1982; Wilson 1983), and possibly an additional intermediate channel (Wilson 1983).

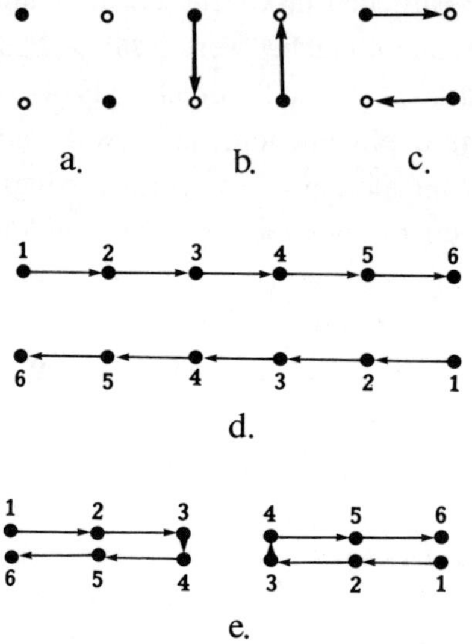


Figure 6.4. The Ambiguous Dot Display. (a) Closed circles appear in the first frame, and open circles appear in the second. The distance between dots is 1° of visual arc. (b) The dots move vertically. (c) The dots move horizontally. (d) The two dots are moved in parallel horizontal paths; the numbers indicate the order of presentation, and arrows indicate the perception of motion when dot positions are separated by 1° of visual arc. (e) The vertical distance between rows is now within the short range limits. Arrows indicate the perceived motion of the dots in this case.

pattern of two dots in motion now occurs at the third and fourth positions). In this case, when the distance between dot positions was large, observers generally perceived the horizontal motion of the dots. There was no ambiguity of movement at the central positions of the dots. When the two horizontal rows were brought closer together, however, the two dots were sometimes perceived as moving toward the center, bouncing off each other, and moving back toward the periphery, as indicated in Figure 6.4e. With a separation between dots within the rows of 1°, this bouncing motion was observed on half of the trials when the separation between rows was roughly 0.6° (this varied slightly with the rate of presentation of the dots). The implication of these results is that when the distance between rows is within the short range spatial limits, the bouncing motion is observed most often. When the separation is small, short range motion measurements signal a vertical motion

**Figure 6.5. The Ternus Configuration.** The **x**'s represent the positions of the dots in the first frame, and the **o**'s represent the positions of the dots in the second.

between the third and fourth positions of the dots, and the final computation of motion may be forced to be consistent with this constraint.

Additional phenomena that demonstrate the powerful constraint imposed by short range motion measurements are those based on the Ternus configuration (Ternus 1926; Pantle and Picciano 1976; Petersik and Pantle 1979). The original stimulus, shown in Figure 6.5, consisted of two frames, each containing two dots. The **x**'s and **o**'s indicate the positions of the dots in the first and second frames, respectively. The rightmost dot in the first frame coincides with the leftmost dot in the second (the experiments of Pantle and Picciano (1976) and Petersik and Pantle (1979) utilized three elements per frame, with two of the elements overlapping in position). Depending on the presentation conditions, two possible perceptions of movement result when the two frames are alternated. In the first, often referred to as *group* motion, the two dots shift back and forth as a group. In the alternate perception, referred to as *element* motion, the dot that is shared between the two frames remains stationary, and the second dot jumps back and forth over the stationary dot. Two factors that influence whether the observer perceives group or element motion are the presentation times for each frame, and the interstimulus interval (ISI). When these temporal parameters are within a range that is suitable for the short range motion process, the observer generally perceives element motion. Pantle and Picciano (1976) found, for example, that with a constant presentation time of 200 msec, an ISI less than about 40 msec yields a perception of element motion for nearly 100% of the observer trials. Under these temporal conditions, short range motion measurements would indicate that the dots that are shared between frames are not moving, and the perception of element motion is consistent with this constraint. The discrete displacement of the outer element in this case may be mediated by a long range process. The perception of motion for the Ternus configuration can therefore be viewed as supporting the idea that short range motion measurements impose powerful constraint on the motion computation.

### 6.1.2. Additional Short Range Demonstrations

In this section, some further perceptual demonstrations are presented, that are similar to previous short range phenomena, in that they yield qualitatively different perceptions of motion, depending on the range of spatial and temporal displacements. We began with the hypothesis that the initial detection of motion takes place at the location of zero-crossings in the output of the $\nabla^2 G$ convolution, and that perceived motion is consistent with at least the sign of $v^{\perp}(s)$ along these zero-crossing contours. We then sought to test this hypothesis by attempting to design stimuli that force a perception of motion that differs from that predicted by the local movement of the zero-crossings. We proceeded by taking advantage of factors that can cause the zero-crossings to move in a direction that differs from the true direction of displacement of the pattern. One of these factors is a sudden change in contrast of a displaced edge in a subsequent frame. This factor underlies Anstis' reversed phi phenomenon, and is used in the first pattern that we present. A second factor is the closer proximity of an element in a subsequent frame that is not the true corresponding element. This is the primary cause for the perceived incoherent motion of dynamic random dot patterns, when displacements exceed the short range limits. Proximity is used to create a particular pattern of zero-crossing movement in the remaining demonstrations presented in this section. This exploration led to some interesting demonstrations, which however are consistent with a motion measurement scheme based on zero-crossings. More broadly, these experiments continue to support the idea that the short range process provides powerful and unyielding constraint for the motion computation.

The first pair of examples is illustrated in Figure 6.6. For each example, two frames of the motion sequence are shown, with one frame above the other. In the actual presentation, the two frames occupied the same area of the display. In the first example, shown in Figure 6.6a, a pattern based on Anstis' reversed phi phenomenon, which in isolation yields perceived motion to the left, is embedded in a larger pattern of bars moving to the right. The first frame consists of a set of black bars on a grey background. In the second frame, the set of bars is displaced uniformly to the right, and the central segment of the middle bars is changed to

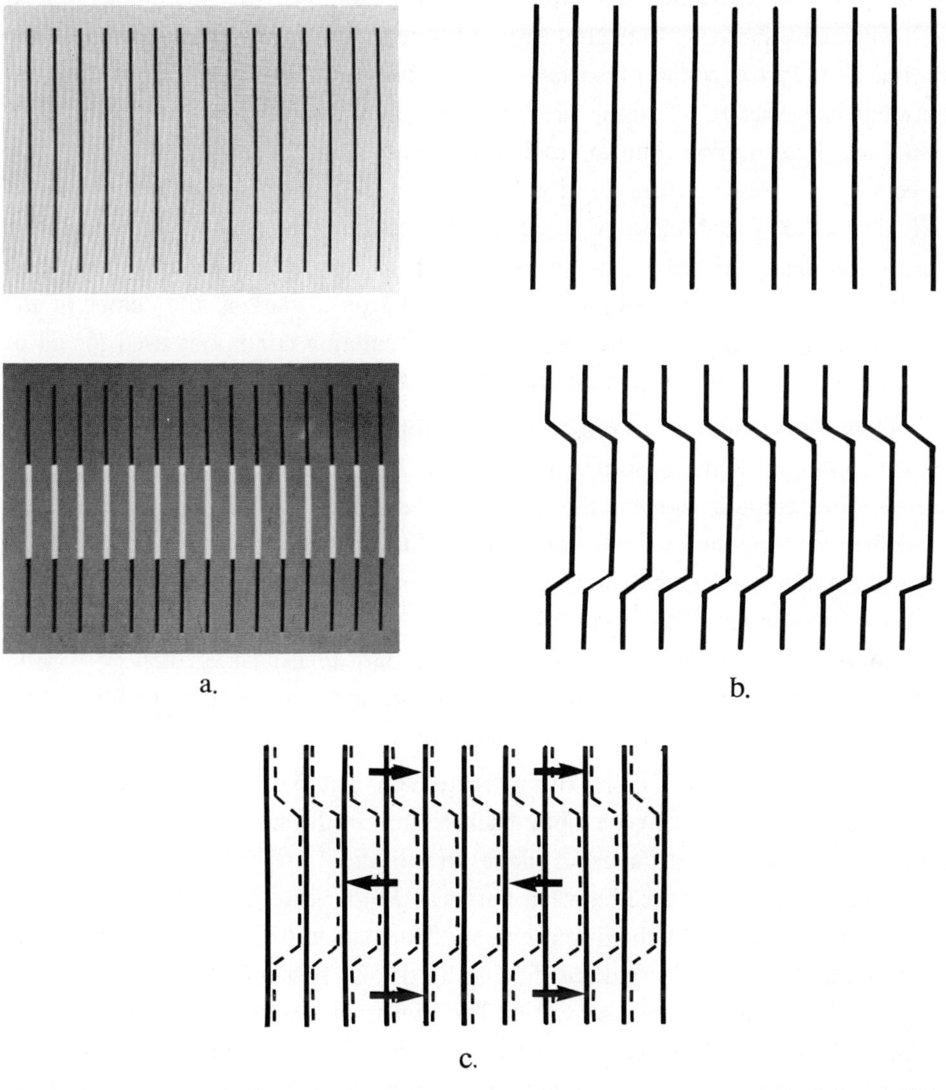

**Figure 6.6. Short Range Motion Demonstrations.** (a) The first frame consists of a pattern of black bars against a grey background. In the second frame, the bars are displaced to the right, and the middle segment of the central bars is changed to white. (b) The second frame consists of bars whose central section has a trapezoid shape. (c) The two frames from (b) are shown superimposed, with dotted lines for the second frame. The arrows indicate the direction of the nearest neighbor in the second frame.

white. The aim of this demonstration was to determine whether the global context of the black bars moving to the right could influence the perceived movement of the central bars. In the actual presentation, barwidth was 2′ of visual arc, the interval between bars was 16′ of visual arc, and displacement between frames was 4′ of visual arc. Presentation time for each frame was 33 msec, and the ISI was a few msec. From a viewing distance of 4 feet, the entire pattern subtended an area of size $4° \times 3°$ of visual angle. A sequence of four frames was displayed repeatedly, creating a steady displacement of the pattern to the right. During presentation of the moving stimulus, fixation was maintained on a mark at the center of the display. (Further details of the display and presentation conditions used for all of the demonstrations in this section are given in Appendix V.)

Under these short range presentation conditions, the central pattern splits from the surround, with the central portion of the bars moving to the left, while the surrounding segments move to the right. This perception is consistent with the local motion of zero-crossings; the larger context of the rightward moving bars, and the vertical continuity of the bars through the center of the pattern do not influence the perceived movement of the central bars. Under long range presentation conditions (for example, a presentation time of 150 msec and an ISI larger than 50 msec), the outer segments of the bars are seen as displacing to the right, while the central segments only flicker between black and white.

The second sequence is shown in Figure 6.6b. The first frame is again a series of vertical bars, now black on a white background. In the second frame, the central portion of the bars has a trapezoid shape. In the actual presentation, barwidth was 2′ of visual arc, the distance between bars was 24′ of visual arc, and displacement between frames (given by the displacement of the bars in the top and bottom thirds of the pattern) was 4′ of visual arc. The entire display subtended $4° \times 3°$ of visual arc, from a viewing distance of 4 feet. The two frames were alternated back and forth, while constant fixation was maintained on a point at the center of the display. Under these conditions, bar segments in the top and bottom of the display have a nearest neighbor in the second frame that is displaced to the right, as shown in Figure 6.6c. For the middle segment of the vertical bars, however, the nearest neighbors in the second frame are displaced to the left. A motion measurement scheme based on zero-crossings predicts a split motion for the pattern, with each segment of the vertical bars moving toward the nearest element in the second frame.

Again, the aim of the demonstration was to determine whether the continuity of the bars in each of the two frames, or the global rightward motion of the entire pattern, could force a perception of movement of the middle segment of the bars that is inconsistent with the movement of the zero-crossings.

The perception of motion of this pattern also differs under short range and long range temporal presentation conditions. With short range parameters (for example, a presentation time of 100 msec and ISI of 10 msec), the pattern splits, exhibiting a strong shearing between the middle and outer sections of the bars. The extent of movement of the outer segments appears smaller, even though the actual displacements are equivalent. One never "sees" the single frame with straight vertical bars; the middle section of the pattern always appears separated from the upper and lower sections. With longer temporal parameters (for example, a presentation time of 100 msec and an ISI larger than 50 msec), the two rigid structures corresponding to the two frames are clearly seen, and the entire pattern moves uniformly in the same direction. In this case, the upper and lower sections of the bars are seen as moving to the nearest element in the next frame, but the middle segments are not. It appears that the continuity of the contours forces the middle segment to a more distant neighbor. It should also be noted that both short range and long range presentation conditions give rise to a strong sense of three-dimensional structure; the middle segment of the bars appears to rise out of the plane of the background.

The perception of motion under short range presentation conditions is again consistent with the local movement of the zero-crossing contours. The continuity of the contours does not force a different perception. It is interesting to consider the perception of motion under long range presentation conditions, in the context of Ullman's theory of the long range motion correspondence computation (Ullman 1979a). Ullman's correspondence scheme is based, in part, on the assumption that distinct elements in the changing image move independently of one another. For the example of Figure 6.6b, if the different segments of the trapezoid-shaped bars were interpreted as distinct elements, then Ullman's minimal mapping scheme, based on the independence hypothesis, would predict a split motion for the pattern, with elements moving to the nearest neighbor in the second frame. The perceived uniform motion of the entire pattern suggests an additional interaction between connected elements along contours, in the long range correspondence computation.

The third example is similar to the previous demonstrations, with the second frame consisting of a set of angle brackets. The two frames are shown separately in Figure 6.7a, and superimposed (with dotted lines for the second frame) in Figure 6.7b. Barwidth was again 2′ of visual arc, and distance between bars in the first frame was 24′ of visual arc. The second frame was constructed in such a way that the distance between the endpoints of the bars in the two frames was 4′ of visual arc. From a viewing distance of 4 feet, the full pattern subtended $4^\circ \times 3^\circ$ of visual arc. For the central part of the pattern, between the dashed lines of Figure 6.7b, the nearest elements in the second frame are displaced to the left. For the outer portions of the pattern, the nearest elements are displaced to the right. The two frames were alternated back and forth. Under short range presentation conditions (for example, a presentation time of 50 msec and ISI of 20 msec), the area of the bars in the vicinity of the outer endpoints clearly moves first to the right, the centermost area of the bars moves first to the left, and the area around the location of the dashed lines in Figure 6.7b is perceived as nonrigid, similar in appearance to a plucked string. From a larger viewing distance, a shearing motion along the dashed lines is seen. The area within the dashed lines splits and moves to the left while the outer areas move right.

Figures 6.7c and d illustrate the movement of zero-crossing contours, predicted by the velocity field algorithm developed in Chapter 4, for each of the two frames. The initial $\nabla^2 G$ operator had a central diameter, $w$, that is half the distance between the bars in the first frame. For the two frames in isolation, the zero-crossings form closed contours, either around the vertical bars (Figure 6.7c) or the angle brackets (Figure 6.7d). A set of intermediate frames was created by linear interpolation of the intensity function between the two frames shown in Figure 6.7a. The initial perpendicular components of velocity were computed from the gradient and temporal derivative of $\nabla^2 G * I$ (where $I$ refers to the particular frame), at the locations of the zero-crossings. Figures 6.7c and d show the resulting velocity fields of least variation, at the two different moments in time at which the two frames are presented. In both cases, the velocity field implies a nonrigid motion, with the outer regions of the bars and brackets moving in a direction opposite the movement of the center. The predicted movement of the zero-crossings is consistent with the perception of movement of the pattern.

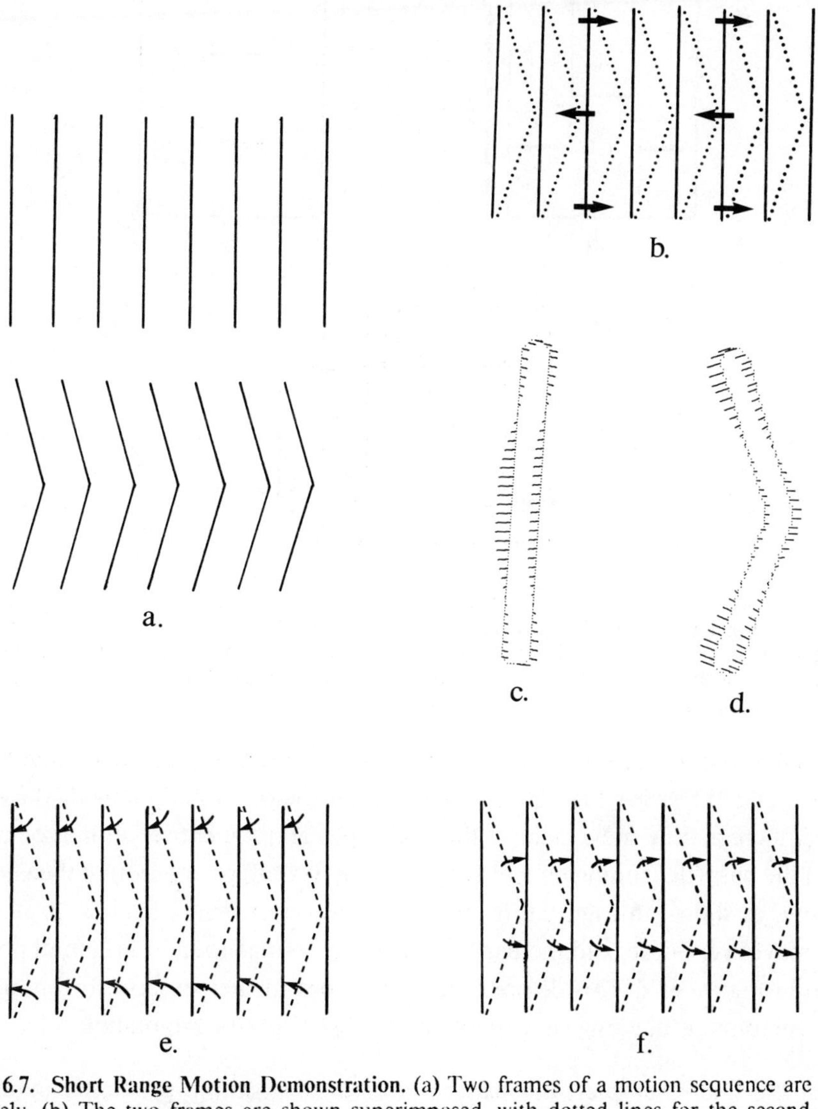


**Figure 6.7. Short Range Motion Demonstration.** (a) Two frames of a motion sequence are shown separately. (b) The two frames are shown superimposed, with dotted lines for the second frame. The arrows indicate the directions of nearest neighbors in the second frame. (c) The predicted motion of the zero-crossings for the vertical bar. (d) The predicted motion of the zero-crossings for the angle bracket. (e) and (f) The perception of motion under long range presentation conditions, for fixation of the endpoints and vertex, respectively.

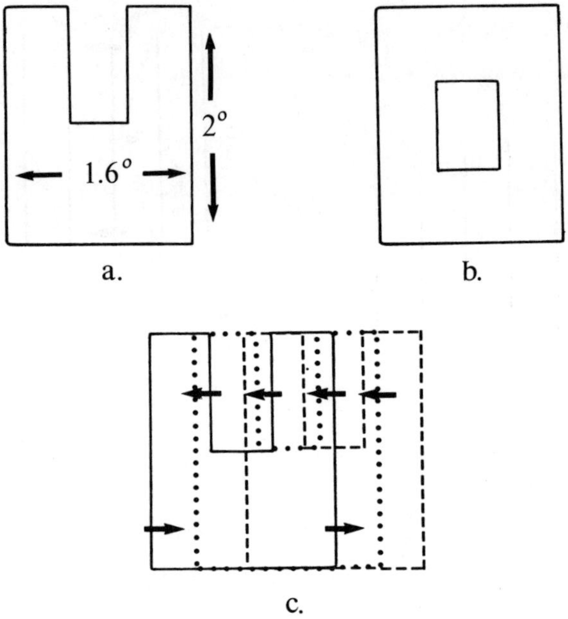


**Figure 6.8. Rigid Figures Undergoing Translation.** (a) and (b) Single frames of motion sequences consisting of simple geometric figures undergoing rigid translation. (c) Three frames are shown superimposed, with dotted and dashed lines for the second and third frames. The arrows indicate the local direction of movement of the zero-crossings.

Under long range presentation conditions (for example, a presentation time of 50 msec and ISI larger than 50 msec), if one of the outer endpoints of the bars is fixated, there is little movement of the endpoints, and the entire top half or bottom half of the bars (depending on which endpoint is fixated) pivots around the endpoint as a unit, as shown in Figure 6.7e. If the central vertex of the angle is fixated, this point moves very little, and the bars are seen as pivoting around the central point, as shown in Figure 6.7f. This demonstration provides further support for interactions along contours in the long range motion correspondence computation.

In the final examples, the pattern consists of a simple geometric figure that is displaced rigidly in the image. Proximity is used to establish a local pattern of zero-crossing movement that differs from the rigid displacement of the figure. The single frames are shown in Figures 6.8a and b, with the dimensions used in the actual presentation. Three frames of the first figure are shown superimposed in Figure 6.8c, using solid, dotted, and dashed lines. True displacement of the figure was 24′ of visual arc to the right. The movie flipped through these three frames, back and forth. The arrows indicate the local direction of movement of the zero-crossings. In the center of the figure, the zero-crossings move toward the nearest neighbor in the next frame, which in this case is displaced to the left. Under short range presentation conditions, and constant fixation on a mark at the center of the display, the perceived motion of the figure is consistent with the movement of the zero-crossings. That is, the figures appear highly nonrigid, with the motion of the central edges out of phase with the motion of the outer edges. Under long range presentation conditions, the figure moves rigidly as a unit.

The results of this last example are consistent with those of Shepard and Judd (1976), who presented observers with the continuous alternation of two perspective views of rigid three-dimensional objects formed by connected blocks. The figures were rotated either in the image or in depth, by angles that ranged from 20° to 180°. Presentation time for the two views was varied, while the ISI was held constant at zero. Shepard and Judd found that the perception of movement of the figures varied with the presentation time. When large presentation times (several hundred msec) were used, observers perceived a rigid rotation of the figure. For intermediate times (60 — 200 msec, depending on the angle of rotation between the two views), movement of the figure appeared nonrigid and incoherent; if the angle of rotation was large, different parts of the object appeared to move independently or to deform into other noncorresponding parts. For very small presentation times (less than 60 msec), the two views appeared superimposed; there was a sense of flicker between the two, but no apparent movement. The times at which the transitions between the three perceptions occurred varied with the angle of rotation between the two views. It is difficult to assess the precise nature of the perception of movement of these figures when the spatial displacements were within the limits of the short range process, because the actual retinal displacements are not described. We would expect, however, that the perception of nonrigid movement at intermediate presentation times was in part the consequence of short range motion measurements that indicate a local direction of movement that differs from the true displacement of the figure. A long range motion correspondence computation, such as that suggested by Ullman (1979a), may also account for some of the nonrigid motions perceived at intermediate presentation times. The perception of movement of the simple objects shown in Figure 6.8, and the more elaborate objects used in the Shepard and Judd experiments continue to demonstrate the important constraint that short range motion measurements impose on the computation of motion.

### 6.1.3. Discriminating Between Alternate Motion Measurement Schemes[3]

In the previous section, we examined a number of short range motion demonstrations, and showed that the perception of movement was consistent with a motion measurement scheme that embodies two computations: (1) the filtering of the visual input with an operator whose spatial weighting function is given by $\nabla^2 G$ (or an approximation, such as the difference of two Gaussians), and whose temporal weighting function is smooth and limited in time, and (2) the detection of the movement of zero-crossings in the output of this filtering stage. For the particular demonstrations that were discussed, the perception of movement could be explained on the basis of properties of the initial spatiotemporal filtering of the visual input. It can be argued that the output of this filtering stage forms the input to the short range motion process, and that the limits on spatial and temporal displacements observed empirically are the consequence of the limited spatial and temporal extent of the initial filtering. From the perspective of the present work, it is important that the movement of zero-crossings in the filtered image be consistent with the perception of movement in these demonstrations, but the zero-crossings are by no means unique in this regard. There are many possible motion measurement schemes that could incorporate the same initial filtering of the visual stimulus, and in principle, account for the existing short range motion phenomena.

In recent years, an interest has developed in the use of psychophysical experimentation for investigating the particular mechanisms by which motion is detected and measured in the human visual system. One question that may be addressed is, at what stage in the analysis of the image does the detection of motion take place? For example, does the detection immediately follow the initial registration of light by the photoreceptors, or is it performed after the subsequent spatiotemporal filtering of the visual input? With regard to the latter, are motion measurements made everywhere in the filtered image, or only in the vicinity of features, such as zero-crossings or extrema (peaks and troughs)? A second question

[3]This section presents material that did not appear in the original Doctoral Dissertation on which this book was based.

that could be considered is, by what mechanism is the detection of motion actually performed? For example, is this measurement achieved through a correlation-type mechanism, as suggested by Hassenstein and Reichardt (1956; Reichardt 1961), a delayed comparison scheme (Barlow and Levick 1965; Burt and Sperling 1981), or via the explicit comparison of spatial and temporal derivatives of the filtered image, as Marr and Ullman (1981) propose? There may not be a single answer to these questions; that is, there may exist multiple pathways in the human visual system with motion detection capabilities, which may perform this detection at different stages in the analysis of the image, using different mechanisms.

In this section, we argue that further discrimination of the particular mechanisms by which motion measurement is performed, on the basis of psychophysical experimentation, requires first, a careful modelling of the early processing stages that precede motion detection, and second, a commitment to a fairly detailed specification of the motion detection mechanism. Aspects of the early processing stages that are important include nonlinearities in the photoreceptor response, and properties of the initial spatiotemporal filtering of the visual input. With regard to the motion detection mechanism itself, there may be some aspects of its implementation that do not critically affect the psychophysical predictions derived from its behavior. For example, suppose that a particular mechanism requires the logical ANDing together of two inputs. Such an operation could also be performed by logically negating one of the inputs (for example, by passing it through an inhibitory interneuron), and applying an AND-NOT operation to the two inputs. These two implementations have different consequences for the neurophysiological investigation of the mechanism, but may yield similar psychophysical predictions. There may be other aspects of the mechanism, however, that critically influence the psychophysical predictions. For example, in the Marr-Ullman model, the mechanism by which the temporal derivative is implemented can significantly influence the behavior of the resulting motion detector.

In the remainder of this section, we attempt to further support the need for detailed models of the early processing stages, and subsequent motion detectors, in order to discriminate between alternate mechanisms for motion measurement, on the basis of psychophysical investigation. The discussion necessarily touches upon some neurophysiological studies as well; the main emphasis, however, is on

**Figure 6.9. The Bulthoff and Gotz Experiment.** This transparent optical wedge filter of linearly graded density was illuminated with a variable light source.

psychophysics. A more detailed account of some of the relevant neurophysiological findings is given in Chapter 8.

### The Need for Modelling the Input Processing Stage

An illusion of motion can originate at various stages in the processing of the visual input; for example, it could be the consequence of nonlinearities in the original photoreceptor response, or a consequence of the initial spatiotemporal filtering of the retinal image. Figure 6.9 illustrates a visual stimulus that produces an illusion for which there is an adequate explanation, based on nonlinear properties of the photoreceptors (Bulthoff and Gotz 1979). (The analysis by Bulthoff and Gotz was performed for a particular class of 'simple' motion detectors, described by Poggio and Reichardt (1976, 1980), which do not require more than two input units in second order nonlinear interaction; this analysis may extend to a broader class of motion detectors as well.) The demonstration was based on previous studies of an illusion of motion created by simultaneously varying the light flux in differently illuminated parts of the visual field (Alpern 1954; Andrews, MacKay and Wilson 1963; Anstis 1967). In the experiments of Bulthoff and Gotz, a transparent optical wedge filter of linearly graded density, shown in Figure 6.9, was illuminated with a variable light source. Rather than perceiving a stationary, temporally-modulating pattern, an observer generally perceived a rotation of the figure toward the less transparent side of the wedge filters during light increase, and rotation in the reverse direction during light decrease. When the light source was modulated, an oscillatory rotation was perceived at low modulation frequencies (0.3 — 2 Hz), a continuous rotation toward the less transparent side of the filters was perceived at high modulation frequencies (5 — 12 Hz), and an ambiguous perception arose at intermediate frequencies (2 — 5 Hz).

One factor that this type of phenomenon has been attributed to is the prolonged latency of the 'light-on' responses of the photoreceptors at low levels of illumination (Lit 1949; for a discussion of the physiological basis for the dependence of latency on illumination level, see, for example, Baylor, Hodgkin and Lamb (1974)). As a consequence of this behavior, there is a delay between the responses of neighboring photoreceptors, for this type of stimulus. This latency, however, is not sufficient to fully explain the illusion (Alpern 1954; Bulthoff and Gotz 1979). The apparent motion has also been attributed to adaptive properties of the input channels of the motion detectors (Andrews, MacKay and Wilson 1963). In a quantitative study of the optomotor behavior of the fruitly, *drosophila melanogaster*, in response to this type of visual stimulus, Bulthoff and Gotz (1979) found that at high modulation frequencies, the dependence of receptor latencies on intensity was adequate to account for the response of continuous motion, assuming the particular class of motion detectors considered. For low frequencies, Bulthoff and Gotz show that the oscillatory response could be explained by the characteristics describing the apparent saturation of photoreceptors in quasi-static conditions (a model of this behavior of the photoreceptors can be found in Dawis (1978)). As a consequence of this second phenomenon, the response of neighboring photoreceptors undergoes a sinusoidal modulation, with the amplitude of modulation varying with the position of the photoreceptors with respect to the stimulus. This again yields an adequate stimulus for the detection of movement by the class of models considered. It is likely that the representation provided by the photoreceptors, in response to this visual stimulus, is also sufficient for the detection of movement by a broader class of motion detection models. To summarize, for this particular motion illusion, it appears that an adequate explanation can be made on the basis of nonlinearities in the photoreceptor response alone (although, as Bulthoff and Gotz note, other explanations may also be possible). In general, to compare the behavior of proposed motion detection mechanisms for a particular visual pattern, properties of the input processing stage must be taken into account, as the results of this stage may alter the predictions of the subsequent motion detectors.

## The Intermediate Spatiotemporal Filtering of the Visual Input

Neurophysiological studies, discussed in Chapter 8, suggest that in the visual systems of higher primates, cells in the retina and lateral geniculate nucleus (LGN) may not exhibit selectivity for the direction of movement of a visual stimulus (with the possible exception of retinal W cells, which may project primarily to the superior colliculus). This suggests that the detection of motion may take place after the processing of the visual input by the retina and LGN. If this is the case, then the nature of this processing is important for the analysis of subsequent motion detection mechanisms. With regard to retinal processing, early electrophysiological studies of retinal ganglion cells (the output cells of the retina) in the cat showed their receptive fields to have an antagonistic center-surround organization (Kuffler 1953), whose shape may be approximated by the difference of two Gaussians (Rodieck and Stone 1965; Rodieck 1965; Enroth-Cugell and Robson 1966). In analytic terms, Rodieck (1965) described the output of the retinal ganglion cells as the convolution of the changing light pattern with the spatial difference-of-Gaussians operator, combined with a particular temporal filtering. The possible role of this processing in the detection of intensity changes was discussed by Marr and Hildreth (1980). The retinal ganglion cells do not form a single homogeneous class; two of the main cell types have been labelled X and Y cells (Enroth-Cugell and Robson 1966; Cleland, Dubin and Levick 1971). In general, X cells exhibit linear summation of their visual inputs, have smaller receptive fields, and exhibit a sustained response to modulations in light intensity. Y cells, on the other hand, are nonlinear, have larger receptive fields, and exhibit transient responses to light modulation. At the LGN, the main properties of the X and Y cells are essentially preserved (Cleland, Dubin and Levick 1971; Hoffman, Stone and Sherman 1972; Dreher and Sanderson 1973). A model of the formation of the spatial and temporal aspects of the X and Y cell responses, via interactions between cell layers in the retina, has been proposed by Richter and Ullman (1980, 1982).

Psychophysical studies also indicate that in the human system, visual information is initially processed through multiple channels that differ in their spatial and temporal response properties (for example, Campbell and Robson 1968; Blakemore and Campbell 1969; Graham and Nachmias 1971; Sachs, Nachmias and Robson 1971; Kulikowski and Tolhurst 1973; Tolhurst 1973, 1975; Spitzberg and Richards 1975; Breitmeyer and Ganz 1977; Cowan 1977; Graham 1977; Watson and

Nachmias 1977; Legge 1978; Wilson and Bergen 1979; Wilson 1980; Burr 1981). Channels tuned to higher spatial frequencies are selective for lower temporal frequencies, and vice versa, making the channels sensitive to different ranges of velocity in the visual input (Fahle and Poggio 1981). High spatial frequency sensitive channels also exhibit a sustained response to modulations in the intensity function, while those sensitive to lower spatial frequencies exhibit transient responses. The sustained and transient channels have been loosely identified with the X and Y cell systems discovered through neurophysiological studies. This dichotomy in the spatiotemporal properties of the channels led some researchers to suggest that the sustained channels are involved primarily in the analysis of the spatial structure of the image, while transient channels play a role in the analysis of movement and temporal change (for example, Tolhurst 1973; Kulikowski and Tolhurst 1973; Ikeda and Wright 1972, 1975). Marr and Ullman (1981), on the other hand, propose a direct interaction between the sustained and transient channels, for the measurement of motion.

Quantitative properties of the spatiotemporal filtering performed by the multiple channels in human vision are important for the analysis of motion detection mechanisms. Through psychophysical studies, however, it is difficult to separate properties of the filtering mechanisms from properties of the motion detection mechanisms themselves. Psychophysical data reveals the response of the system as a whole, to a particular visual stimulus, rather than isolating the response of a particular stage in the visual pathway. It is nonetheless important that the predictions of a computational model of the combined filtering and motion measurement stages be consistent with the results of psychophysical studies.

Several quantitative models of the channel properties have been proposed. As we mentioned earlier in this chapter, with regard to the spatial properties of the filtering performed by these channels, Wilson and Bergen (1980) present one model, based on psychophysical studies of threshold spatial vision, that proposes the existence of at least four channels, whose underlying spatial response function may be approximated in shape by the difference of two Gaussians. For the higher spatial frequency (sustained) channels, the spatial functions are modelled as ideal, balanced difference-of-Gaussian operators (that is, they exhibit little or no response to uniform illumination), with a ratio of space constants for the two Gaussians of roughly 1:1.75. The spatial functions underlying the lower spatial frequency (transient) channels are progressively less balanced, due to weaker inhibition by the larger negative Gaussian relative to the excitation of the central, narrower Gaussian. The ratio of space constants for the transient channels are modelled as 1:3.0. The frequency bandwidths at half height for these spatial functions are roughly 1.33 octaves for the sustained channels, increasing to about 2.0 octaves for the transient channels (Wilson 1983). Other researchers have proposed spatial filters with narrower bandwidths (Sachs, Nachmias and Robson 1971; Kulikowski and King-Smith 1973; Watson 1980) and broader bandwidths (Spitzberg and Richards 1975). In the central fovea, the width of the central positive regions of the spatial difference-of-Gaussians operators is 3′, 6′, 11′, and 21′ for the four channels in the Wilson and Bergen model. The size of these operators increases linearly with eccentricity. Recent studies of suprathreshold spatial vision have led to a modification of the Wilson and Bergen model, incorporating two additional channels in the fovea (Wilson 1983), one with a spatial response function of smaller size, and the other of intermediate size. Other perceptual studies also support the existence of an additional smaller operator in the central fovea, sensitive to higher spatial frequencies (Marr, Poggio and Hildreth 1980; Watson 1982).

A second quantitative model of the initial spatial filtering has been proposed by Watson (1983). In this model, the spatial response functions are described by Gabor functions (the product of a sinusoid and a Gaussian), with a frequency bandwidth of roughly one octave. The model includes eight initial operators, whose spatial frequencies (of the underlying sinusoid) range in one octave intervals from 0.25 to 32 cycles/degree. In both the Wilson and Bergen, and Watson models, the filters are assumed to be oriented. In Watson's model, the filters have an orientation bandwidth of about 38°, and their primary orientations are evenly spaced at intervals of 36°, covering the full range of image orientations with five oriented filters.

Psychophysical studies have also attempted to measure the temporal properties of the filters underlying the multiple channels (for example, Watson and Nachmias 1977; Breitmeyer and Ganz 1977; Legge 1978; Wilson 1980). Watson and Nachmias (1977) hypothesized that the spatial and temporal properties of these filters were separable; that is, the overall spatiotemporal response could be expressed as the product of a spatial response function and temporal response function. Direct evidence in support of this hypothesis, for the case of the transient channels, was presented by Wilson (1980). The psychophysically measured temporal response function for the transient channels appears to be composed of a positive phase of about 60 msec, followed by a negative phase. Such a function was suggested by Watson and Nachmias (1977), and is supported by the data of Tolhurst (1975), Breitmeyer and Ganz (1977), and Legge (1978). The negative phase may be somewhat longer than the positive one, or may be followed by damped oscillation of small amplitude (Breitmeyer and Ganz 1977). Marr and Ullman (1981) proposed that the output of the transient channels represents the temporal derivative of the spatially filtered intensity function. The computation of a derivative requires a temporal impulse response function consisting of positive and negative phases of roughly the same shape. Marr and Ullman (1981) demonstrated a close agreement between the predictions of their computational model, and the predicted response of the transient U-channel in Wilson's (1980) model, for lines and bars moving with a retinal velocity of 3°/second. The temporal response function for the sustained channels consists mainly of a single, positive phase that rises quickly, and then falls off slowly in amplitude over time, with a primary temporal integration period of roughly 200 msec (Tolhurst 1975; Breitmeyer and Ganz 1977; Legge 1978).

It should be apparent that the precise quantitative properties of the spatial and temporal filtering performed by the early stages in the visual system are not fully known. This complicates the analysis of the predictions of motion detection mechanisms, which either incorporate this filtering as part of their mechanism, or operate on the output of this filtering stage. Some motion phenomena may be accounted for solely on the basis of this spatiotemporal filtering; any motion detectors incorporating this filtering may then yield similar behavior. Thus, together with the input processing stage discussed earlier, properties of this intermediate filtering stage should also be taken into account, in formulating the predictions of subsequent motion detectors.

## Modelling the Motion Detection Stage

Several mechanisms have been proposed for the detection of motion in biological systems. Four examples are illustrated in Figure 6.10. These particular detectors respond preferentially to motion from left to right; complementary mechanisms are required to detect motion in the opposite direction. Figure 6.10a shows the mechanism proposed by Hassenstein and Reichardt (1956; Reichardt 1961) for the visual system of the fly. The two inputs, $\mathbf{I}_1$ and $\mathbf{I}_2$, are multiplied after low-pass

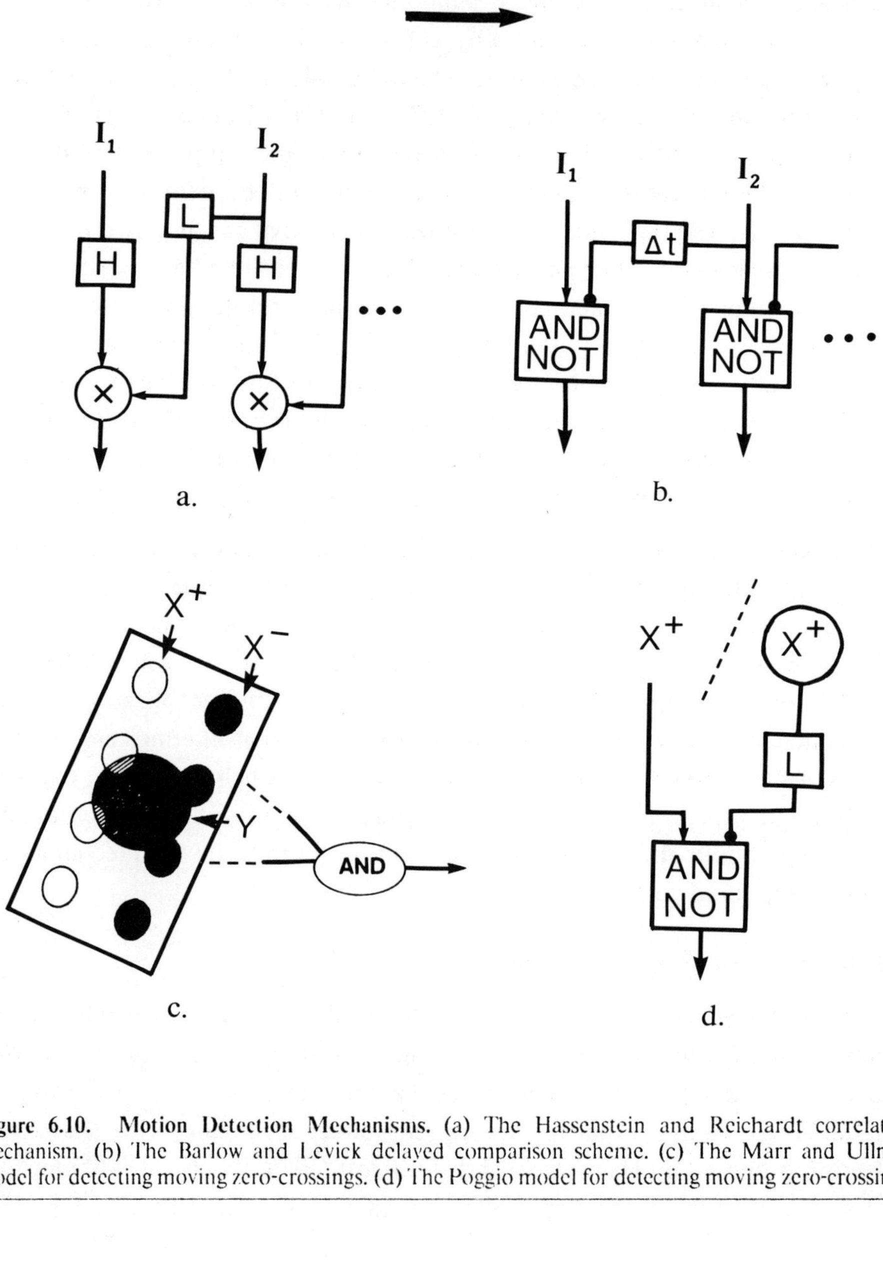


**Figure 6.10. Motion Detection Mechanisms.** (a) The Hassenstein and Reichardt correlation mechanism. (b) The Barlow and Levick delayed comparison scheme. (c) The Marr and Ullman model for detecting moving zero-crossings. (d) The Poggio model for detecting moving zero-crossings.

filtering with different time constants. If an averaging operation is performed on the output, the overall operation of this mechanism is equivalent to the cross-correlation of the two inputs. Figure 6.10b shows the mechanism proposed by Barlow and Levick (1965) to account for the directional selectivity of ganglion cells in the rabbit retina. An AND-NOT operation is applied to the input $\mathbf{I}_1$, and a delayed version of the input $\mathbf{I}_2$. (A similar mechanism is described by Burt and Sperling (1976).) The delay $\Delta t$ may also be replaced by a low-pass filtering operation. A synaptic mechanism for the neural implementation of the AND-NOT or multiplication operations of the Hassenstein-Reichardt and Barlow-Levick models has been proposed by Torre and Poggio (1978). This first set of mechanisms belong to a general class of motion detection models that are essentially equivalent in their steady state response (Poggio and Reichardt 1976, 1980).

The predicted response of a motion detector to changing visual input depends on the stage of processing at which the detection is performed. The inputs to the above motion detection mechanisms could, in principle, represent the output of any of the early processing stages. For example, these inputs might represent the photoreceptor response, as suggested by Hassenstein and Reichardt (1956) for the visual system of the fly, or the results of an intermediate retinal processing stage, as suggested by studies of directional selectivity in the rabbit retina (Barlow and Levick 1965). In higher primates, where directional selectivity is a common property of cortical neurons, the ascending fibers from the LGN may provide the input to motion detection mechanisms.

Psychophysical studies by Movshon *et al.* (1983) suggest that the first stage in the measurement of motion by the human visual system involves an orientation-selective spatial analysis, and provides only the component of velocity in the direction perpendicular to the orientation of spatial features in the image. Orientation selectivity is embodied in the motion detection models of Marr and Ullman (1981) and Poggio (1983), in which the motion detection is combined with the detection of oriented zero-crossing segments. The Marr-Ullman (1981) model, illustrated in Figure 6.10c, is based on the explicit comparison of spatial and temporal gradients of the $\nabla^2 G$-filtered image, at the locations of zero-crossings. The model was proposed specifically for a class of simple cells in primate visual cortex, that exhibit selectivity for edges in their receptive field of a particular orientation, sign of contrast, and direction of motion. The $\mathbf{X}^+$ and $\mathbf{X}^-$ inputs in Figure 6.10c represent the output of the ON-center and OFF-center X cells ascending from the LGN. The **Y** input represents the output of either an ON-center or OFF-center geniculate Y cell. In the original model, these inputs, arranged in the spatial configuration shown, are followed by an AND operation. As we mentioned previously, Marr and Ullman assume that the X cell system carries the convolution of the visual pattern with the $\nabla^2 G$ operator (or its approximation, the difference of two Gaussians), and propose that the Y cell system carries the temporal derivative of this convolution. The spatial arrangement of the $\mathbf{X}^+$ and $\mathbf{X}^-$ inputs provides the motion detector with selectivity for zero-crossings of a particular orientation and sign of contrast (Marr and Hildreth 1980), and the additional **Y** input renders the detector selective for the direction of motion of a zero-crossing contour segment. The mechanism for detecting moving zero-crossings, suggested by Poggio (1983), is shown in Figure 6.10d. An AND-NOT operation is applied to adjacent X cell inputs of similar type, and the use of an explicit temporal derivative is replaced by low-pass temporal filtering of one of the inputs. The resulting motion detector represents a hybrid between the mechanisms shown in Figures 6.10a through c. The Poggio model is supported in part by the experiments of Schiller (1982), which indicate that the selective blocking of the ON-center system in the retina has little affect on the orientation and direction selectivity of cortical neurons.

In order to distinguish between the alternate mechanisms for motion detection shown in Figure 6.10, on the basis of psychophysics, a commitment must be made to a fairly detailed specification of the mechanisms. With regard to zero-crossing based motion detectors, there are several neural mechanisms by which the detection of zero-crossings, and measurement of direction of movement, may be implemented. The nature of the implementation of these two aspects of the detector can influence the subsequent psychophysical predictions. For example, in the Marr-Ullman model, the presence of a zero-crossing is inferred by the existence of two adjacent peaks (or areas of significant amplitude) of opposite sign, in the $\nabla^2 G$-filtered image. Such a mechanism should, in principle, only respond when there is a true zero-crossing present. In the Poggio model, shown in Figure 6.10d, the zero-crossing is inferred by the presence of a peak (or area of significant amplitude) of one sign, adjacent to an area in which there is no significant signal of that sign. The use of an ideal AND-NOT operation yields a detector that responds both to true zero-crossings, and to isolated peaks in the $\nabla^2 G$-filtered image. It may be possible to design a

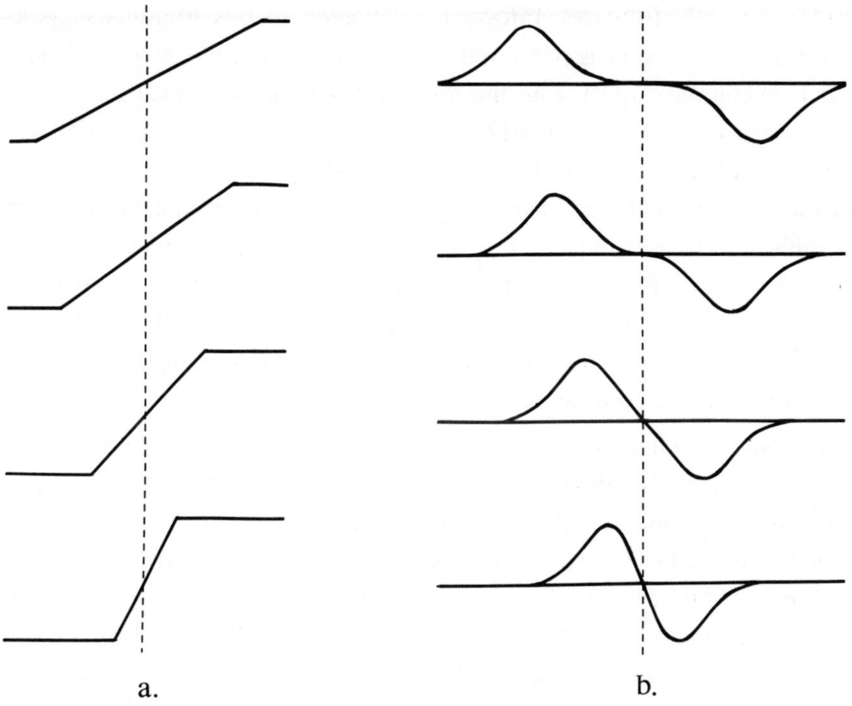


**Figure 6.11. A Sequence of Ramp Profiles.** (a) A sequence of intensity profiles, consisting of a single ramp-shaped intensity change, whose width decreases with time. (b) The convolution of the profiles in (a) with the one-dimensional $D^2G$ operator.

similar detector that does not respond to isolated peaks (Kobi Richter, Tomaso Poggio, personal communication).

Figure 6.11 illustrates a perceptual demonstration for which the distinction between the detection of zero-crossings, and isolated peaks, is relevant. The visual stimulus consists of a temporal sequence of bars, whose intensity profiles (cross-section perpendicular to the orientation of the bars) are given by the sequence shown in Figure 6.11a. Each profile consists of a ramp-shaped intensity change, with the width of the ramp changing over time. Figure 6.11b shows the convolution of these profiles with the one-dimensional $D^2G$ operator. The theoretical zero-crossing always occurs at the center of the ramp; if the centers of the ramps are aligned in space, then in theory, the zero-crossing never moves. The positions of the peaks in the $D^2G$-filtered profiles change with the width of the ramp, however, approaching its center as width decreases. Informal observation of this stimulus suggests that motion toward the center is perceived, consistent with the movement of the peaks. This demonstration suggests that the movement of the theoretical zero-crossing in an ideally filtered image is not sufficient to account for the perception of motion.[4] A movement detector such as that suggested by Poggio (1983), however, may yield a motion measurement that is more consistent with perceptual behavior, due to its possible sensitivity to isolated peaks in the $\nabla^2 G$-filtered image. It is also the case that while the theoretical zero-crossing has an instantaneous spatial location, any neural zero-crossing detector will be active within a small neighborhood of the zero-crossing, rather than providing a motion measurement only when the zero-crossing is centered in its receptive field.

The issue of whether zero-crossings, extrema (peaks and troughs), or other features of the $\nabla^2 G$-filtered image form the primitive elements at which subsequent computations take place has also arisen in studies of other visual tasks. For example, psychophysical studies suggest that peaks may be explicitly detected and matched in stereopsis (Mayhew and Frisby 1981). Studies of the spatial localization of edges indicates a close agreement between perceived location, and the location of zero-crossings (Watt and Morgan 1982). Subsequent studies of the localization and description of different types of edges may suggest that peaks play an important role in this visual task (Watt and Morgan 1983).

For feature-based motion detectors, the mechanism by which movement is detected may also influence subsequent psychophysical predictions. For example, consider the explicit use of the temporal derivative of the $\nabla^2 G$-filtered image, as in the Marr-Ullman model. Suppose that a zero-crossing were present at a particular location in the image, and the intensity function around the location of the zero-crossing is changed over time, in a way that maintains the zero-crossing at a constant position (the sequence of profiles in Figure 6.11a represents an example of such a modulation). Although the zero-crossing is stationary, the modulation of intensity over time may yield a nonzero temporal derivative, which is then likely to yield a measurement of movement by a Marr-Ullman type of motion detector.

[4] It is expected that the nonlinear properties of the photoreceptors and intermediate spatiotemporal filtering will alter the behavior of the zero-crossing in this case, but these additional factors may not be sufficient to account for the perception of movement, if an ideal theoretical model of zero-crossing detection is assumed.

The perception of movement in such a visual stimulus may be consistent with the behavior of a neural mechanism that is designed to detect moving zero-crossings, even though the theoretical zero-crossing is actually stationary. On the other hand, it may be possible to design a motion detector that only signals movement when there is physical displacement of a zero-crossing contour. This type of mechanism may not predict a perception of movement in a temporally modulated intensity function, for which the zero-crossing remains in a constant spatial location.

**Summary**

To summarize, a theoretical model of the computations underlying the detection and measurement of motion in the human visual system may be implemented in neural hardware in many different ways. In order to distinguish between the use of different computational models, on the basis of psychophysical experimentation, it is necessary to consider aspects of this implementation. Factors that are important include nonlinearities in the initial photoreceptor response, properties of the intermediate spatiotemporal filtering of the image, and quantitative aspects of the motion detection mechanism itself. All of these factors can influence subsequent psychophysical predictions.

## 6.2. The Smoothness Assumption

In this section, we address the second of our main questions concerning the measurement of motion in the human visual system: *Is the perception of motion consistent with the computation of the velocity field of least variation?* Several of the examples of smooth curves in rotation, examined in Section 4.4.1, were chosen because perceptual studies suggest that human observers perceive a motion that differs from the true movement of the curves. It was found that the smoothest velocity field also differs from the true velocity field for these examples. In Section 6.2.1, the smoothest velocity field is compared directly with the perception of motion, for these and other motion demonstrations. The comparisons are qualitative, because the relevant perceptual studies are largely phenomenological. The development of a quantitative means for assessing the perception of the velocity field is an important area for future work.

In Section 3.3.4, some alternate formulations of the smoothness constraint were described briefly. In the first, smoothness of the velocity field was imposed

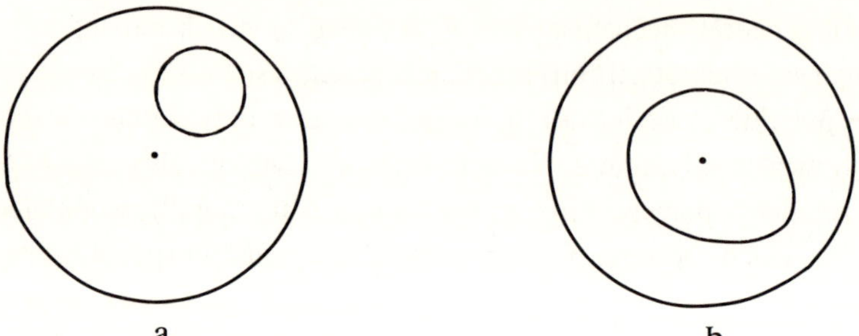


Figure 6.12. A Rotating Circle. (a) A circle is placed eccentrically on a rotating turntable. (b) A deformed circle is placed centrally on the rotating turntable, and observed at its center.

over two dimensions, rather than along one-dimensional contours. In a very different formulation, this constraint was expressed in terms of the deformation of contours over time, rather than the variation of velocity along contours. These alternate formulations raise questions for the psychophysical investigation of motion measurement, which are discussed in Section 6.2.2.

### 6.2.1. Examples of the Perception of the Smoothest Velocity Field

When a homogeneous circle is placed eccentrically on a rotating turntable, as shown in Figure 6.12a, the circle itself does not appear to rotate; rather, it remains upright during the course of its rotation (Wallach, Weisz and Adams 1956). In other words, the circle is seen to undergo pure translation, even when its background clearly rotates. In Section 3.3.3, it was noted that if there exists a valid solution to the velocity field computation that corresponds to pure translation, then this solution minimizes $\int |\frac{\partial \mathbf{V}}{\partial s}|^2 ds$. Thus, the smoothest velocity field for the rotating circle corresponds to pure translation, consistent with the perception.

Suppose the circle is then deformed, placed centrally on the turntable, and the observer fixates its center as the turntable rotates (Figure 6.12b). Again, observers do not perceive a rigid rotation of the figure. Rather, a continuous deformation takes place; the body of the circle appears to remain still while the deformation travels about its perimeter (Wallach, Weisz and Adams 1956). The rotating deformed circle of Figure 4.6 was based on this demonstration. The smoothest velocity field was found to exhibit significant movement around the area of the bulge, with little motion of the circle further away. This velocity field is again qualitatively consistent with observers' perception of the deformed circle.

Wallach, Weisz and Adams (1956) also observed the motion of rotating ellipses with various aspect ratios. When an ellipse whose axes measured 25 and 23.5 cm was rotated about its center, it appeared to stand still while its contour pulsated. The largest effects were observed for an ellipse whose aspect ratio was 3:2. In this case, the entire figure appeared fluid, undergoing a strong deformation, as well as a rotation. For some of the observers, the deformation was more restricted, and did not occur in the immediate vicinity of the poles of the major axis of the ellipse. A very narrow ellipse appeared more rigid. Wallach, Weisz and Adams found the motion of the ellipse to be the same, regardless of whether it rotated about its center, or was placed eccentrically on the turntable. They also noted that if the ellipse was first observed at rest, it required from 5 seconds to a minute before naive observers reported the figure as fluid, whereas all of the observers immediately reported the figure as fluid when the turntable was already rotating at the beginning of the observation period.

In Figure 4.4, the smoothest velocity field was illustrated for an ellipse whose aspect ratio was 2:1. This velocity field clearly implies a significant distortion of the contour, in addition to a rotation. In the immediate vicinity of the poles of the major axes, the smoothest velocity field is very similar to the true velocity field. In Figure 6.13, the true and smoothest velocity fields are shown for rotating ellipses whose aspect ratios are 25:23.5 and 5:1.[5] When the ellipse is nearly circular, the smoothest velocity field indicates a strong inward and outward pulsation of the contour (Figure 6.13b). The smoothest velocity field for the narrower ellipse, shown in Figure 6.13d, is closer to the true velocity field than in the case of an ellipse with aspect ratio 2:1 (see Figure 4.4c), implying less distortion of the contour. There is, however, still some implied distortion of the contour at this aspect ratio. Finally, the smoothest velocity field for an eccentrically rotated ellipse differs from that of the centrally rotated ellipse only by the addition of a uniform translation along the contour. Thus, it is expected that the same deformations of the ellipses will be implied by the smoothest velocity field obtained for eccentric rotations. We conclude that

[5]The angular velocities of the ellipses in Figures 4.4 and 6.13 differ; if we let the angular velocity of the ellipse in Figure 4.4 be denoted by $\omega$, then the ellipses of Figures 6.13a and c are rotated with angular velocities of roughly $\frac{3}{2}\omega$ and $\frac{1}{2}\omega$, respectively.

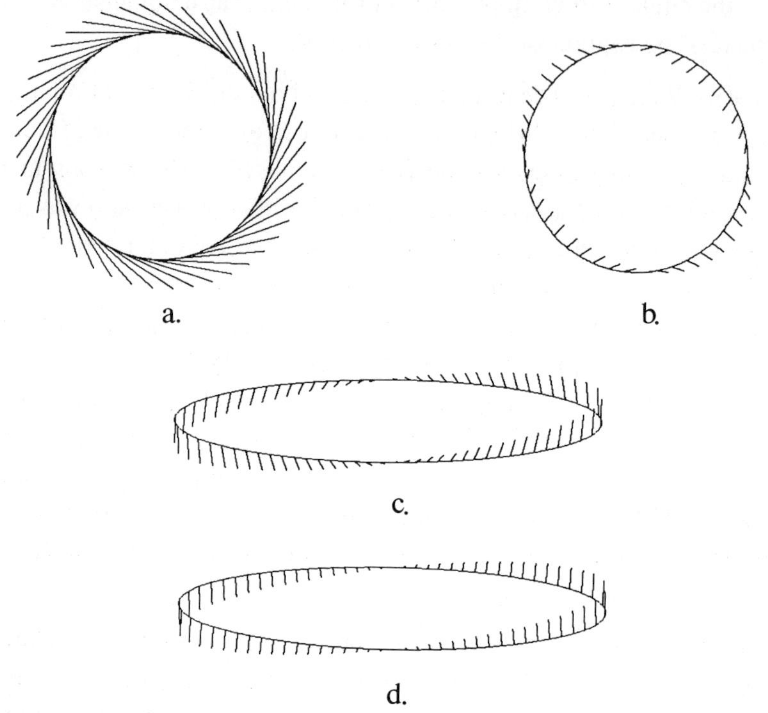


**Figure 6.13. Rotating Ellipses.** (a) The true velocity field for a rotating ellipse, whose aspect ratio is 25:23.5. (b) The computed velocity field of least variation for the ellipse in (a). (c) The true velocity field for a rotating ellipse whose aspect ratio is 5:1. (d) The computed velocity field of least variation for the ellipse in (c).

the perception of movement of rotating ellipses is qualitatively consistent with the computation of the velocity field of least variation. It is not yet clear, however, why a significant time period is required before nonrigid motion is perceived, when the ellipse is first observed at rest.

It is well known that a spiral appears to expand or contract, when it undergoes pure rotation about its center (Holland 1965). The perceived velocity field thus contains a large radial component, while the true velocity field contains only a rotational component of velocity. Figure 4.5 illustrated the smoothest velocity field for a single arm of a rotating logarithmic spiral; this velocity field exhibits a large radial component of motion at the center of the spiral, which decreases toward the periphery. The perception of movement of the spiral is qualitatively more consistent with the smoothest velocity field, than with either the true velocity field, or the initial perpendicular velocity vectors shown in Figure 4.5b.

The ability of the human visual system to interpret the three-dimensional structure of objects from their projected two-dimensional motion was first explored systematically by Wallach and O'Connell (1953) in a study of what they termed the *kinetic depth effect.* In their experiments, an unfamiliar wireframe object was rotated behind a translucent screen, and the shadow of its parallel projection was viewed from the other side of the screen. Given the projection of the wireframe figure shown in Figure 4.7a, for example, observers were able to give a correct description of the hidden object's three-dimensional structure from its two-dimensional projection. This particular figure is an example of an object for which it can be proven that the true and smoothest velocity fields are equivalent; this equivalence was demonstrated in Figure 4.7. If the analysis of motion in the human visual system is a two-stage process, with the measurement of two-dimensional motion preceding the interpretation of three-dimensional structure from motion, then the perception of the correct three-dimensional structure of this figure would suggest that the human system also computes the correct two-dimensional motion initially.

Other smooth curves in rotation that were examined in Section 4.4.1 also appear to yield a perceived motion that is consistent with the smoothest velocity field. For example, if the rotating helix is viewed while maintaining fixation on a mark at the center of the helix, a downward movement is perceived, consistent with the smoothest velocity field of Figure 4.8d. The smooth curve of Figure 4.10, rotated about its vertical axis, appears to deform slightly toward the top and bottom areas of the contour, consistent with the velocity fields of Figures 4.10d, 4.11d and 4.12d. When rotated about the axis $\mathbf{n} = [1, 1, 1]^T$, as shown in Figure 4.13a, the curve first appears to translate as a unit toward the upper right corner of the display. At other parts of the curve's trajectory (for example, those given by the curves of Figures 4.14 and 4.15), the three-dimensional structure of the curve is more apparent, suggesting that a velocity field is derived at these parts of the trajectory that more closely resembles the true velocity field. (Details of the presentation of these smooth curves in rotation are given in Appendix V.)

Another motion demonstration of interest was presented by Anstis and Ramachandran (Anstis 1983). The first frame of a motion pair consisted of a

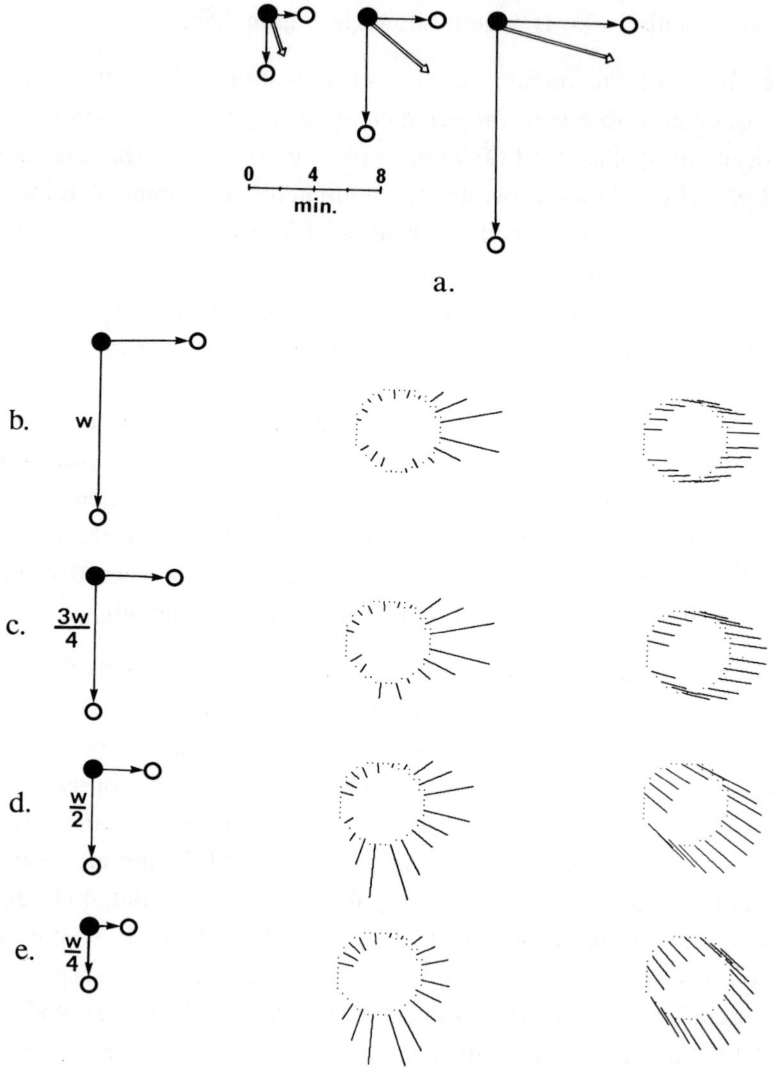


**Figure 6.14. The Dot Demonstration.** (a) Closed circles represent a dot in the first frame, and open circles represent dots in the second. The double arrows show the perceived direction of motion of the full dot pattern. The inserted scale indicates the size of displacements. (b) through (e) The leftmost column indicates the displacements of dots as a function of $w$, the middle column illustrates the initial perpendicular components of velocity, and the rightmost column illustrates the computed velocity field of least variation.

sparse array of random dots. In the second frame, each of the dots was replaced by two dots, with one displaced horizontally, and the other vertically. The local configuration of dots is shown in Figure 6.14a. The filled circle represents a dot from the first frame, and open circles represent dots in the second frame. The horizontal displacement was always half the vertical displacement. The inserted scale indicates the size of the displacements used in the actual presentation, in minutes of visual arc. Anstis and Ramachandran found that when the two frames were alternated, the entire field of dots appeared to move obliquely as a unit. The perceived direction of velocity varied with the absolute displacement of the dots. Viewed closely (with large displacements), perceived direction of velocity was roughly horizontal. As viewing distance increased (displacements decreased), the apparent direction of motion became increasingly more vertical.

The predictions of the velocity field algorithm, based on the smoothness constraint, are shown in Figures 6.14b through e. In the leftmost column, the actual displacements between dots are given as a function of $w$, the diameter of the central positive region of the initial $\nabla^2 G$ operator. The initial perpendicular components of velocity, shown enlarged in the center column, were computed by comparing the temporal derivative and gradient of $\nabla^2 G * I$ (where $I$ is the image). The resulting smoothest velocity fields are shown enlarged in the rightmost column of Figures 6.14b through e. As the absolute displacement of the dots decreases, the predicted direction of motion becomes increasingly more vertical, consistent with the perception.

Adelson and Movshon (1983) present a motion demonstration, referred to as the "split herringbone." The display, shown in Figure 6.15a, consists of alternating columns of line segments that tilt left or right in the odd and even columns. The odd (left-tilting) columns move upward, while the even (right-tilting) columns move downward. There are two perceptions of motion that result, depending on the viewing conditions. When the display is of high contrast, sharp, and centrally fixated, the columns are seen moving vertically, in opposite directions. When the display is of low contrast, blurred with a diffusion screen, or viewed peripherally, an illusion of rightward motion is seen. To explain this phenomenon, Adelson and Movshon (1983) distinguish two kinds of velocity field computations. The first is a local computation, that involves the tracking of localizable features, such as the line endpoints in the herringbone pattern. The second is a more "global" process that combines velocity constraints derived from different parts of the image. They argue that the perception of vertically moving columns represents the results of the first type of process, while the perception of rightward movement represents the results of the second.

It is possible that both perceptions arise from a single underlying velocity field computation that combines velocity constraints along contours in the image, which is applied to image descriptions derived from initial $\nabla^2 G$ operators of different size. Figures 6.15b and c show the zero-crossing contours derived from the convolution of the image of Figure 6.15a with $\nabla^2 G$ operators whose central positive diameter, $w$, is 6 and 16 picture elements, respectively. With $w = 6$, the zero-crossing contours surround individual bars, while in the case where $w = 16$, the zero-crossings merge from one column to the next, forming continuous contours across the image. Figures 6.15d and e illustrate the initial perpendicular components of velocity, obtained from the temporal derivative and gradient of $\nabla^2 G * I$. Figures 6.15f and g illustrate the resulting smoothest velocity field in each case. For the smaller operator, the velocity directions are not strictly vertical, but deviate by $7° - 9°$ from the vertical direction. If strictly vertical movement is perceived, it could either be due to less error in the initial measurements of $v^{\perp}(s)$, or to an additional tracking of the endpoints of the lines, as Adelson and Movshon suggest. For the larger operator, the smoothest velocity field indicates horizontal movement of the contours. Horizontal motion was perceived under the conditions of peripheral viewing or blurring, for which a coarser initial filtering of the image is expected. Rightward movement was also seen in low contrast displays; this may be due to a greater sensitivity of low spatial frequency channels at low contrast.

When surfaces undergo rotation, it is possible that the perceived velocity field varies with the speed of rotation. For all of the examples of moving curves shown here and in Chapter 4, we computed the velocity field of least variation consistent with the instantaneous motion measurements derived from the changing image. If the speed of rotation is high, the visual system may not have sufficient time to compute the velocity field of least variation; rather, it may only be able to compute a velocity field that is intermediate between the initial and smoothest velocity fields. If this is the case, then figures in rotation may appear less rigid as the speed of rotation increases.

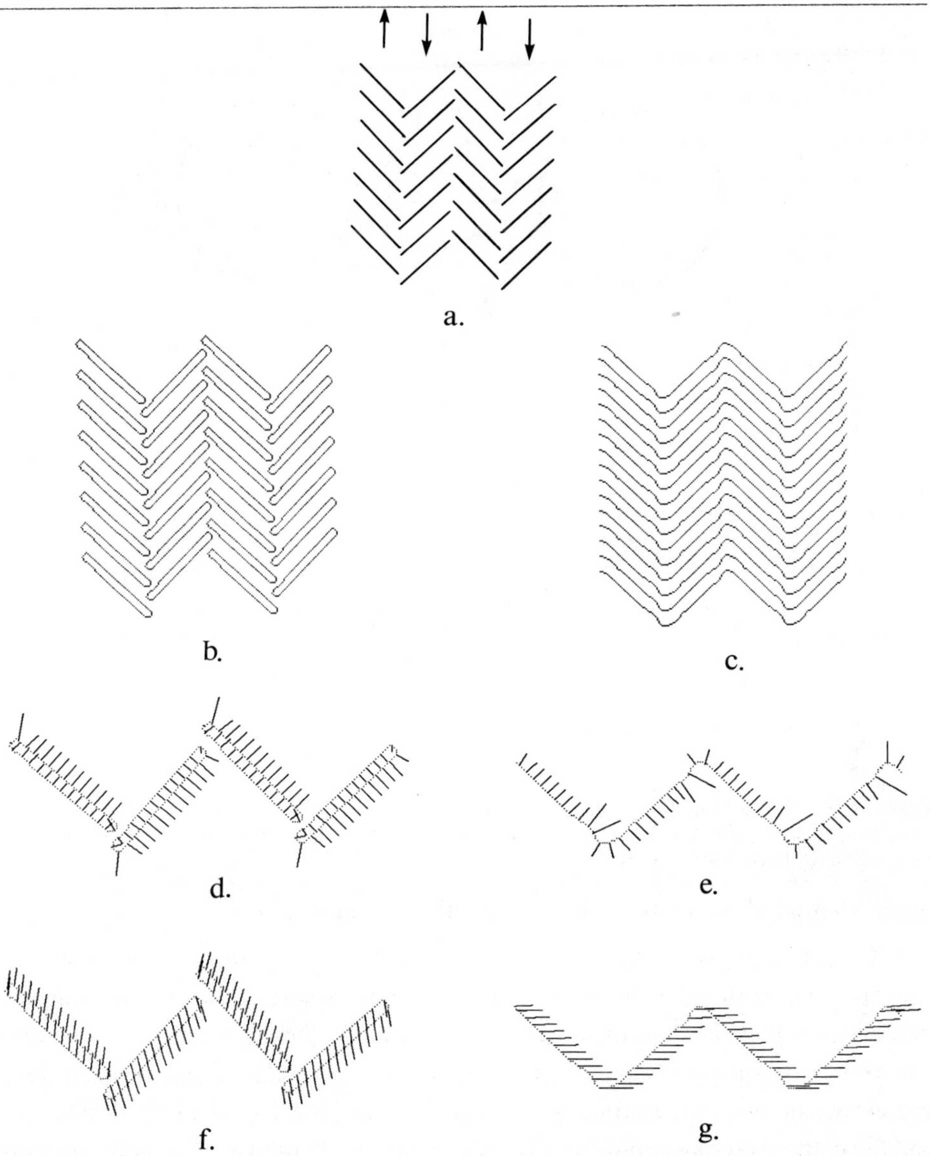


**Figure 6.15. The Split Herringbone Demonstration.** (a) Alternating columns of a herringbone pattern of lines move in opposite directions. (b) and (c) The zero-crossing contours derived from the convolution of the herringbone pattern with $\nabla^2 G$ operators of size $w = 6$ and 16 picture elements, respectively. (d) and (e) The initial perpendicular components of velocity. (f) and (g) The computed velocity fields of least variation.

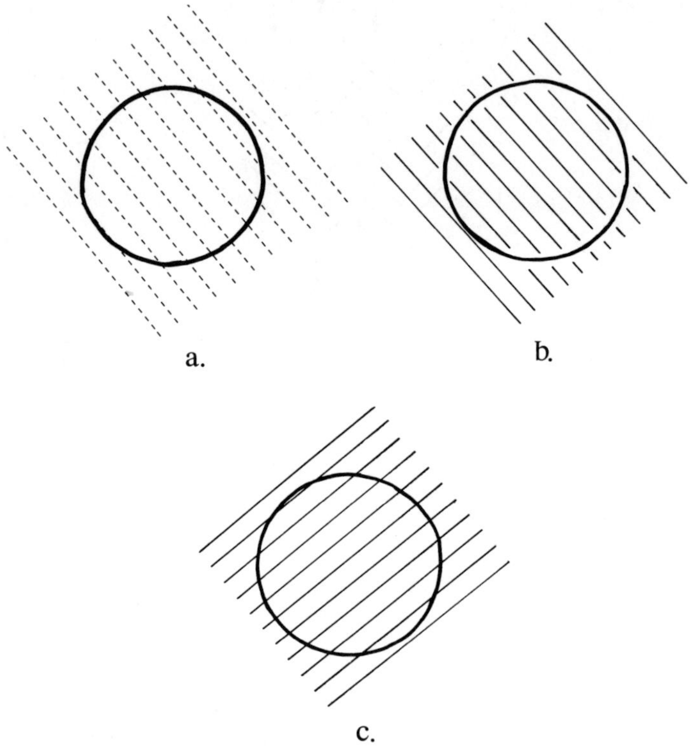

Figure 6.16. Integrating Constraints Along Contours. (a) The background lines are of lower contrast than the circle. (b) The lines and circle do not intersect. (c) The lines and circle intersect, and appear with the same constrast.

### 6.2.2. Alternate Formulations of the Smoothness Constraint

The smoothness constraint has been formulated in terms of the variation in velocity along contours in the image. One question that this raises for psychophysical investigation is whether velocity constraints from the image are in fact integrated along contours, such as the zero-crossing contours, or over small areas. Wallach (1976) presented a motion demonstration in which a line was drawn on a textured wallpaper, and the entire pattern was moved vertically downward behind a rectangular aperture. Even with a clear vertical movement of the background, the line appeared to split from the background and move horizontally (its perceived direction of motion without the background). It was noted earlier that a circle placed on a rotating turntable appears to remain upright, even when the background rotation of the turntable is clear. We explored some variations on this experiment, shown in Figure 6.16. These patterns were drawn on white paper, and rotated peripherally on a turntable. The figures were then viewed with fixation maintained at the center of the turntable. If the background lines in the pattern are either of different contrast (indicated by the dotted lines), or are disconnected from the circle, as in Figures 6.16a and b, the circle splits from the background and remains roughly upright as the turntable rotates. That is, the circle appears to translate as the background rotates. If the background is of sufficient contrast, and intersects the circle, the circle appears to rotate. The result is similar for patterns placed at the center of the turntable. For the patterns of Figures 6.16a and b, the zero-crossing contours for the lines and circle are disjoint, while in the pattern of Figure 6.16c, the contours from the lines and circle merge together. If constraints on the velocity field were integrated over small areas of the image, then the proximity of the background lines and circle in Figures 6.16a and b should be sufficient to yield a perception of rotation of the circle. The fact that the circle appears to remain upright in these demonstrations suggests that perceived motion may be more consistent with the results of a comparison that integrates velocity constraints along connected contours in the image.

If smoothness of the velocity field is imposed along contours in the image, and the measurement of motion involves the computation of the velocity field of least variation, one can then ask, which measure of variation is used for this computation? In Section 3.3.2 and in Appendix II, it was shown that a measure of variation based only on one component of velocity, such as direction, in general does not yield a unique velocity field solution. It may be the case, however, that further constraint (beyond smoothness) is used to identify a unique solution. Such a computation may, for example, yield a velocity field for which there is less variation in direction than in the velocity field that results from minimizing the measure of variation given by $\int |\frac{\partial \mathbf{V}}{\partial s}|^2 ds$. Before a comparison of this type can be performed, however, alternative computations based on different measures of variation must be formulated in detail.

In Section 3.3.4, we briefly mentioned a formulation of the smoothness constraint in terms of the deformation of contours over time, rather than the variation of velocity along contours. A computation that attempts to minimize local deformation over time requires the explicit use of subsequent positions of the moving curve as a source of constraint on local motion. In Section 3.4, one method by which an additional curve could provide constraint for the velocity field computation was illustrated. This approach was not pursued further, in part because there are indications from perceptual studies that a second curve may not be used as an explicit source of constraint in the human visual system. One piece of evidence is derived from a study by Diener *et al.* (1976) of the perceived speed of spatial patterns of light and dark bars. In the case of periodic patterns, if constant fixation is maintained at a single location on the display, the perceived speed of patterns moving at constant velocity increases with spatial frequency. That is, a high frequency pattern is perceived as moving faster than a low frequency pattern, when both patterns are actually moving with the same absolute speed (also see Burr and Ross (1982)). This effect does not require periodic patterns; if a single bar is moved across the visual field during constant fixation, a narrower bar is perceived as moving faster than a wide bar. Contours occur at the edge of each bar in these patterns, and move with its absolute speed. If a subsequent position of the contours is used explicitly to constrain the measurement of their movement, then perceived speed is expected to be proportional to absolute speed, rather than to vary with the spatial frequency of the pattern.

The explicit use of a second curve for constraining local velocity, in the way described in Section 3.4, assumes that velocity is constant over the time interval separating the two frames. Another piece of evidence suggesting that a second curve may not be used in this way is derived from the perception of an expanding and accelerating circle. A movie of an expanding circle was created with a set of discrete frames, in which the radius of the circle was doubled between frames. (Details are given in Appendix V.) A velocity field computation that both tries to minimize variation in velocity, and force the velocity field to project to a subsequent curve would predict a rotational component of velocity, together with the radial component, as shown in Figure 3.8; however, the circle is perceived as undergoing pure radial motion, and expanding at an accelerated rate. This and the previous experiment suggest that the motion computation in the human visual system may not explicitly use a subsequent position of a curve to constrain the velocity field, but further investigation is necessary to resolve this question.

There is a different way in which subsequent positions of a curve may be used, mentioned briefly in Section 4.5.2. It is possible to compute the full two-dimensional velocity field from a single curve and the instantaneous measurements of $v^{\perp}(s)$

along the curve. It may be the case, however, that at each moment, only a small number of iterations of the computation are performed, and the current velocity field is then passed on to the curve at the next moment in time (forming the initial velocity field at the next time), where further iterations are performed. In this way, the velocity field is built up over a larger time frame, covering multiple positions of the curve (Horn and Schunck 1981). This type of computation yields a velocity field that varies smoothly over time as well as space. A velocity field computation that is extended in time has two possible implications for the perception of motion. First, two frames presented in a brief flash may not be sufficient to compute a highly varying velocity field, such as that generated by a complex smooth curve rotating in space. Second, one would expect to find an effect of visual momentum, as suggested by Ramachandran and Anstis (1983; Anstis 1983); that is, an abrupt change in the direction of motion of an object should not be perceived as such. With regard to visual momentum, however, it may be difficult to distinguish between a smoothing due to the underlying temporal properties of the initial filters, and that due to an extended velocity field computation.

### 6.3. Deriving Further Motion Constraints From Eye Movements

In this work, we have considered the computation of the velocity field, in the absence of eye movements. This allows greater control over the retinal input to the computation, so that direct comparisons can be made between the performance of the algorithm, and human motion perception. Both saccadic and smooth pursuit eye movements, however, are likely to play an important role in the measurement of motion. We expect that a detailed velocity field computation, such as that discussed in this book, only takes place in the central part of the visual field. One role of eye movements may then be to bring an area of interest in the peripheral part of the visual field within the central visual field for detailed analysis. One affect that these eye movements will then have on the local velocity field computation is to alter its retinal input, by adding a translation and possibly a small rotation to the incoming image. There may, however, be several other ways in which eye movements influence the analysis of motion. In this section, we briefly describe one means by which smooth pursuit eye movements could provide additional constraint for the velocity field computation.

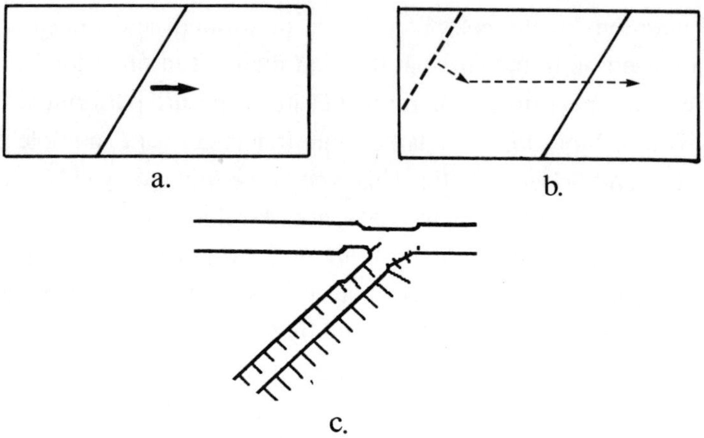


**Figure 6.17. Wallach's Aperture Experiment.** (a) An oblique line moves behind a rectangular aperture. (b) The perceived direction of motion of the line, at two different positions in its trajectory. (c) The local zero-crossing contours, together with the perpendicular velocity vectors.

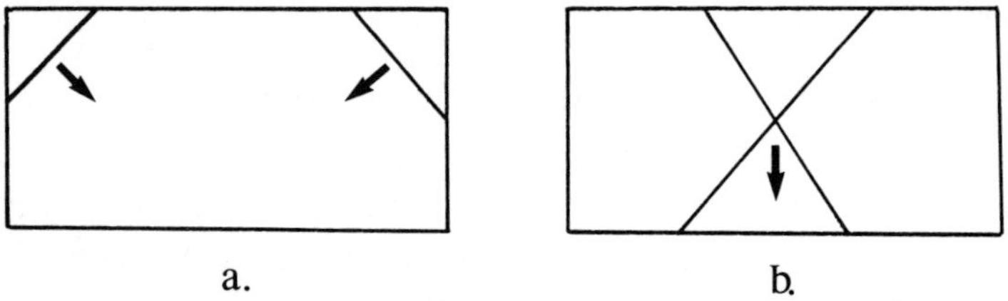


**Figure 6.18. Wallach's Crossed Lines Demonstration.** (a) Two lines first enter the aperture from the upper left and right corners. (b) When the lines meet, their moving configuration is also consistent with a rigid cross moving downwards.

Suppose there is a feature in the image, whose movement is ambiguous. If the feature is then tracked with a smooth pursuit eye movement, the feature may unambiguously take on a velocity corresponding roughly to the movement of the eye. One example of this phenomenon is given by the Wallach experiment shown in Figure 6.17. A line oriented 45° from the horizontal is moved behind a rectangular aperture; its perceived motion, with a fixed eye position, is indicated in Figure 6.17b. As the line enters the upper left corner, it appears to move in the direction perpendicular to its orientation, but when it meets the lower left corner, it changes direction and moves horizontally. If the intersection of the line with the horizontal

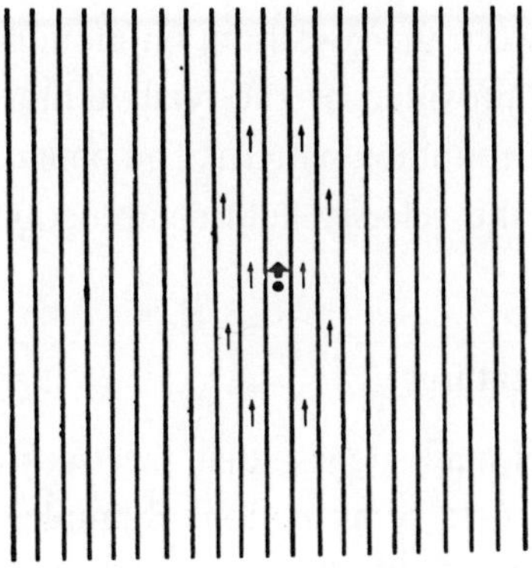

**Figure 6.19. The Schulman Experiment.** The dot is tracked vertically, while the lines remain stationary.

or vertical border of the aperture is tracked, the line is perceived as moving in the same direction as the eyes. A second example, also from Wallach (1976), is shown in Figure 6.18. Here, two lines enter the aperture from the upper left and right corners (Figure 6.18a), then meet toward the center of the aperture and cross over each other (Figure 6.18b). Throughout its trajectory, the local pattern is also consistent with two lines forming a rigid cross, moving downwards. With a fixed eye position, the lines are perceived as separate, and sliding across each other, even if the lines begin their motion in the crossed configuration. If the intersection of the lines is tracked, however, a rigid cross moving downwards is perceived.

In both of the above examples, it is possible that the image feature being tracked is assumed to correspond to the same physical feature under motion, and it therefore may take on a velocity direction that is equal to the direction of movement of the eyes. In this way, tracking eye movements could provide additional constraint for the velocity field computation. Features in the image, such as the lines in the Wallach demonstration, which remain in roughly the same retinal location during a smooth pursuit eye movement, may also take on a velocity corresponding to the motion of the eyes. This is suggested by an experiment by Schulman (1979), illustrated in Figure 6.19. If the subjects tracked the vertically moving dot, while the vertical lines remained stationary, the lines appeared to move vertically as well. While tracking eye movements appear to provide a useful source of constraint for the velocity field computation, the nature of this constraint may be, in general, quite complex. A study by Wallach, O'Leary and McMahon (1982), for example, reveals a nontrivial interaction between the constraint on motion provided by tracking eye movements, and constraint provided by the relative movement between an object in motion, and a moving surrounding pattern. The potential role of eye movements in providing constraint for the velocity field computation can be explored further through perceptual studies.

### 6.4. Tracking Localizable Features

Perceptual experiments may suggest that the tracking of localizable features plays a role in the velocity field computation. Examples include Wallach's (1976) experiments with lines moving behind apertures, shown in Figures 6.17 and 6.18. These demonstrations suggest that the movement of the intersection of the line with the aperture may constrain its direction of motion. The perceived vertical movement of the columns of the herringbone pattern studied by Adelson and Movshon (1983) may also indicate a tracking of line endpoints. The explicit use of localizable features can be examined more closely, both from a perceptual and computational viewpoint.

With regard to perception, one can first ask whether the presence of moving features (such as endpoints, breaks, dots, local deformities, or discontinuities in line orientation) can influence the perception of motion. If this is the case, it can then be asked whether the tracking of features is a low-level, automatic process that occurs everywhere in the image, or a process that requires either tracking eye movements or attention. The Adelson and Movshon (1983) experiments, and some of Wallach's (1976) demonstrations were performed with steady fixation, suggesting that eye movements alone cannot account for the phenomena. However, the experiments do not rule out the possibility that the use of features requires that we explicitly attend to their movement.

From a computational standpoint, one can first ask whether the explicit detection and tracking of localizable features is necessary, to account for the perceptual phenomena. For a feature such as the intersection of the line and aperture in Figure 6.17a, its movement along the aperture may be given directly by the integration of $v^{\perp}(s)$ along contours. The local zero-crossing contours around the intersection are shown in Figure 6.17c, along with the perpendicular velocity vectors. Along the horizontal aperture, $v^{\perp}(s) = 0$; the local combination of constraints therefore yields a horizontal movement of the line. It is possible that, in general, the measurements of $v^{\perp}(s)$ alone are sufficient to account for the influence of the movement of special features. If this is not the case, these features must be unambiguously identified at two moments in time, and then tracked. These two computations are nontrivial. First, these features may be straightforward to identify in the stylized stimuli used in perceptual experiments, but are difficult to define in a natural image. Lawton (1983) presents one scheme for locating *distinctive* points along zero-crossing contours, which are usually points of high curvature. Once identified, the tracking process itself may be nontrivial, if the density of features is high; it amounts to solving the correspondence problem (Ullman 1979a). In Chapter 4, the incorporation of constraint from tracked features into the velocity field computation was shown to be straightforward; the algorithm then computes the velocity field of least variation that is consistent with both $v^{\perp}(s)$ along a contour, and the known velocity points. The tracking of features may contribute to the velocity field computation, but the integration of multiple measurements of $v^{\perp}(s)$, and use of additional constraint such as smoothness, are still required.

### 6.5. Short Range and Long Range Motion Processes

This book has addressed the measurement of motion by a short range process; that is, one that requires approximately continuous motion. We mentioned in Chapter 2 that there may also exist a long range process in human vision, which is capable of inferring motion between stimuli presented with much larger spatial and temporal intervals (the long range process may also operate at shorter ranges, as suggested by Mather and Anstis (1983)). The possible existence of these two processes raises questions concerning the role they play in motion measurement, and how they interact with one another. Perceptual studies indicate that at some stage during the analysis of motion, the short range and long range processes interact in some way (Clatworthy and Frisby 1973; Green and Von Grunau 1983). In Section 6.1, we pointed out that many of the short range phenomena indicate that short range measurements impose very powerful constraint on the motion computation, which cannot easily be overridden, if at all, by the long range process (a point also made by Mather and Anstis (1983)).

For some visual tasks, such as the detection of a sudden movement, or object segmentation, rough and instantaneous motion measurements alone may be sufficient (Marr and Ullman 1981). Computational studies of the recovery of three-dimensional structure suggest that a more complete and accurate measurement of motion is required for this task (Ullman 1979a, 1980, 1983a,b; Prazdny 1980, 1983; Clocksin 1980; Longuet-Higgins and Prazdny 1981; Longuet-Higgins 1981). The work presented in this book suggests that a computation based on short range motion measurements, in principle, can yield a very precise local velocity field. The reliable derivation of three-dimensional structure, however, requires that these measurements be integrated over a large time frame (Ullman 1983). The long range process may play a role in this integration process. Studies by Petersik (1980) suggest that the long range process may be crucial to the recovery of structure from motion in the human visual system. A second possible role for the long range process is the maintenance of continuity of objects moving through a complex environment, in which objects are sometimes occluded from view (Marr 1982, Braddick 1983); for example, when an animal runs through a forest. This task could also be guided by information supplied by the short range motion process. The general picture that emerges is one expressed by Ullman (1981); that the short range and long range motion processes may interact in complementary ways to produce a flexible and reliable system for motion measurement. Both computational and perceptual studies are needed, to explore this possibility further.

### 6.6. The Use of Multiple Channels for Motion Measurement

In Section 6.3.1, the spatial and temporal properties of the initial filtering of the retinal image were discussed. Recent psychophysical experiments by Adelson and Movshon (1982, 1983) have addressed the interaction of multiple spatial frequency channels for motion analysis. In these experiments, stimulus patterns consist of two superimposed sinewave gratings, at different orientations, moving in the direction perpendicular to their orientation. A schematic diagram of a typical stimulus is shown in Figure 6.20a. A vertical grating moving to the left is added to an oblique grating moving down and to the right. Together, the two gratings form a single rigid pattern, moving in a direction consistent with the constraints imposed by the two individual components. That is, the movement of each grating constrains the velocity of the combined pattern to lie along a line in velocity space, as shown in Figure 6.20b, and the intersection of these lines indicates the motion of the pattern. Under some conditions, the gratings do not form a single coherent pattern perceptually; rather, they appear to split and move independently of one another.

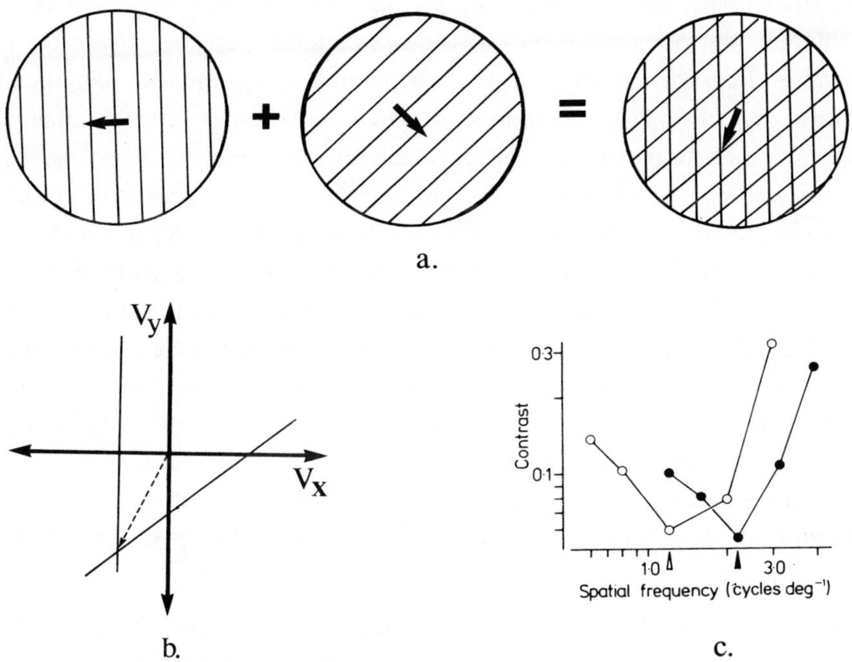


**Figure 6.20. The Adelson and Movshon Experiment.** (a) Two sine wave gratings, shown schematically, are superimposed to form a single rigid pattern moving down and to the left. (b) The intersection of lines of constraint in velocity space. (c) The contrast of the second grating that is required to see coherence on half the trials is plotted as a function of spatial frequency (from Adelson and Movshon 1982).

In Adelson and Movshon's experiments, the spatial frequency, orientation, and contrast of the individual gratings was varied, and subjects were asked to judge the coherence of the resulting pattern.

Coherence of the combined pattern was found to decrease with an increase in any of the following factors: (1) the difference in contrast of the component gratings, (2) the angle between their primary directions, (3) the difference between their spatial frequencies, and (4) their speed of movement. The effects of contrast and spatial frequency interact. The data from one of Adelson and Movshon's experiments is reproduced in Figure 6.20c. The contrast of one grating was fixed at 0.3, and its spatial frequency was fixed at 1.2 cycles/degree (open circles) and 2.2 cycles/degree (closed circles). The two plots indicate the contrast of the second grating that was required for the subjects to see coherent motion on 50% of the trials, as a function of the spatial frequency of the second grating. For a given contrast difference, as the difference in spatial frequency of the two patterns increases, a split motion is seen more often. This spatial frequency dependence suggests that components of the moving pattern that fall within the spatiotemporal range of a single channel are combined, while those that fall within separate channels are not. Psychophysical studies discussed in Section 6.1.3 indicate that the channels in human vision may be quite broadly tuned for spatial frequency, resulting in significant overlap in the frequency sensitivities of the individual channels (Wilson and Bergen 1979; Spitzberg and Richards 1975). For the patterns utilized by Adelson and Movshon, gratings that differ significantly in spatial frequency (ie, those that differ by at least one octave) may provide primary stimulation for separate channels. If the combination of motion constraints only takes place within single channels, the resulting pattern may appear incoherent, with the two gratings moving independently of one another. Due to the broad frequency tuning of the channels, however, two gratings with significantly different spatial frequency may be combined if their contrast ratio is appropriate.

It is not surprising that visual information analyzed by different channels in human vision may be treated independently, with regard to the processing of motion information. Their differing spatiotemporal properties render them optimally sensitive to different ranges of velocity, so that information available from the high spatial frequency (low temporal frequency) channels may be quite different from that available from the low spatial frequency (high temporal frequency) channels. In many natural scenes, objects move over a wide range of velocities. Suppose, for example, that an observer is moving in a car, through an environment in which objects appear over a large range of depths. Objects in the foreground tend to move with very high velocities, while mountains at a distance tend to move very slowly with respect to the observer. No single channel can optimally measure motion over the full range of velocities that the human visual system typically encounters.

Suppose that the full set of channels in the human visual system were used in the measurement of motion. If a pattern moves across the retina with sufficient speed that only the low spatial frequency channels are capable of analyzing its motion correctly, the coarse spatial resolution of these channels would result in only a rough spatial analysis of the pattern. Velocity information from these coarser channels may be used, however, to direct appropriate eye tracking, which could bring the retinal motion of the pattern within a range that can be analyzed by the higher spatial frequency channels. This then allows a higher precision in the velocity measurement, and a more detailed spatial analysis of the pattern. This possible role of multiple channels in motion analysis is similar to the role proposed by Marr and Poggio (1979) for multiple channels in stereopsis. In the stereo algorithm proposed by Marr and Poggio, low spatial frequency channels allow fusion to take place over a larger range of disparities; through the control of vergence eye movements, disparity information from these channels is used to bring the left and right images into a range of disparities that the higher spatial frequency channels may fuse.[6] This allows for the fusion of a large range of disparities, with high resolution of disparity information. An analogous interaction between multiple spatial frequencies channels could be proposed for the analysis of motion. Mutch and Thompson (1982) present an algorithm that utilizes information from low spatial frequency channels to resolve the ambiguities that arise from the measurement of motion of elements in the output of high spatial frequency channels.

### 6.7. The Independence of Motion Measurement and Interpretation

We have assumed in this work that the analysis of motion may be considered a two-stage process, with the measurement of two-dimensional motion in the changing image taking place prior to and independent of the interpretation of the three-dimensional structure of surfaces from their motion. This assumption underlies most computational work on the analysis of motion; proposed methods for motion measurement are generally aimed at computing the two-dimensional motion of elements in the image, and methods for deriving three-dimensional structure have assumed that the projected two-dimensional motion of surfaces is known. The separation of the two stages has been made in computational studies, in part because it simplifies the theoretical analysis of motion measurement and interpretation. The main theoretical problem for both stages is the inherent ambiguity of motion and structure, from information in the changing image. There are many ways of assigning a two-dimensional motion to elements in the image, and many ways of attributing a three-dimensional structure to the underlying surfaces, on the basis of the projected two-dimensional motions. In this book, a method has been presented for measuring

[6]Mayhew and Frisby (1981) suggest a different role for multiple channels in stereopsis, based on the results of psychophysical studies.

the projected motion of general surfaces, in a way that yields a unique solution. Theoretical work on the derivation of three-dimensional structure from motion has largely addressed the requirements for establishing a unique interpretation, under varying assumptions (Ullman 1979a, 1980, 1983a,b; Clocksin 1980; Longuet-Higgins and Prazdny 1980; Prazdny 1980, 1983; Longuet-Higgins 1981; Tsai and Huang 1981; Hoffman and Flinchbaugh 1982; Bobick 1983). This theoretical work indicates that if the analysis of motion is divided into two parts, then each part on its own may be performed in a way that is computationally sound. It may be very difficult, if not impossible, to derive a unique three-dimensional motion and structure directly from the changing image, in a way that is not decomposable into two independent stages. Further theoretical work is needed, however, before this point can be established conclusively.

The possible separation of motion analysis into two stages may also be examined through perceptual studies. One question that can be addressed is the extent to which other three-dimensional cues can influence the measurement of motion. Ullman (1979a) presented some perceptual demonstrations that indicate that three-dimensional cues derived from perspective line drawings do not influence the two-dimensional nature of the long range motion correspondence process. With regard to the short range motion process, one could examine whether three-dimensional cues derived from perspective, stereo, shading, or contour information, influence the perception of the projected two-dimensional motion of continuously moving objects. It is likely that these additional cues influence the interpreted three-dimensional structure of an object in motion (Ullman 1983b). One may find, however, that this influence never yields an interpretation that corresponds to a different projected two-dimensional motion. If this were the case, it would lend support to the view that the measurement of motion is essentially an independent, low-level, two-dimensional computation, which is not influenced by processes involved in the recovery of three-dimensional structure. Such a result would also demonstrate that two-dimensional motion imposes powerful constraint on the recovery of three-dimensional structure.

Utilizing the present work, one could attempt to interface the velocity field computation with a structure from motion computation, and examine whether the computed three-dimensional structure of a moving contour is consistent with its perceived structure. This analysis requires a method for computing three-dimensional structure that is not restricted to deriving only rigid interpretations. Recently, Ullman

(1983b) proposed one such method, that can cope with considerable deviations from rigidity. The algorithm maintains an internal model of the viewed object, and modifies the model at each instant, by the minimal nonrigid change that is sufficient to account for observed transformations in the image. The algorithm presently requires an explicit correspondence between features over time. The integration of the measurement and interpretation stages would be interesting, because there are many examples of smooth three-dimensional curves in rotation, for which the velocity field computation described in this book yields a solution that differs from the true projected two-dimensional motion. If the three-dimensional interpretation is then forced to be consistent with the computed velocity field, the computed three-dimensional structure would be expected to embody deformations of the contour, consistent with the deviations in two-dimensional motion. It would be important that the perceived structure of the moving contour also incorporates these deformations. Preliminary observations of the rotating smooth contours discussed in Sections 4.4.1 and 6.2.1 suggest that this may be the case; that is, rigid curves in rotation often appear nonrigid, in a way that is expected from the smoothest velocity field. Such experimentation can only test for consistency between the predictions of a particular motion computation and human perception; it does not provide proof that the analysis of motion takes place in two stages. Unfortunately, alternative formulations of the computations are not presently available for direct comparison. Independent of particular computational models, it would be useful to establish a quantitative means of assessing the perceived motion and structure of an object, so that rigorous comparisons with the predictions of models can be made.

# 7. THE PSYCHOPHYSICS OF DISCONTINUITY DETECTION

In this chapter, we discuss some psychophysical studies that are relevant to the computations underlying the detection of motion discontinuities in the human visual system. The primary experimental tool that has been used in these studies is the Julesz' random dot pattern, which allows the removal of all other visual cues to the location of surface boundaries, leaving the motion cue alone. The primary interest in this study is the ability of the human visual system to localize the boundaries along which motion is discontinuous. A distinction can be made between this visual task, and the simple judgement that relative movement exists between two areas of the image. The ability of the human visual system to perform these two tasks appears to be quite different. To study the localization of motion discontinuities, an experiment was designed that requires the discrimination of the shape of a boundary between two adjacent areas undergoing different motions in a random dot pattern.

Previous work on the detection of motion discontinuities in random dot patterns is discussed in Section 7.1. In Section 7.2, we examine the dependence of boundary localization on the difference in direction and magnitude of velocity across a boundary. A key question for the algorithm used by the human visual system to localize motion discontinuities concerns the input representation for the computation. Two possible input representations are the initial measurements of the perpendicular components of velocity along contours in the image, and the full two-dimensional velocity field (other representations are also possible). In Section 7.3, a visual pattern is presented, for which the velocity field is difficult to describe (except in the immediate vicinity of fixation), but in which motion discontinuities are very vivid, over an extended area. This demonstration is used to suggest that the detection of motion discontinuities may begin to take place prior to the combination of initial motion measurements for the full two-dimensional velocity field computation.

It was noted in Chapter 5 that in the case of discrete presentation, a trade-off exists between the range of spatial displacements that can be tolerated for the detection of coherent motion, and the resolution with which discontinuities may be analyzed. This trade-off is a consequence of the initial processing of the image with a range of operators that differ in their spatial and temporal properties, and can be observed psychophysically. In Section 7.4, the results of an experiment are presented that suggest that the size of a differentially moving patch in a random dot pattern, which is required for discrimination against its background, increases with the size of the discrete displacement of the patch.

### 7.1. Contributions of Previous Work

The first important psychophysical result was simply that the human visual system is capable of detecting motion discontinuities, and using them to infer surface boundaries (Anstis 1970; Regan and Spekreijse 1970; Julesz 1971). Braddick's studies (1973, 1974, 1980) show that there is a limit on the spatial displacement and time interval required to perceive coherent motion of a dense random dot pattern. Coherent motion of two adjacent surfaces is not essential to perceive their boundary, however; it is possible to detect the boundary between coherent and incoherent fields of motion as well. Later work (Baker and Braddick 1982a) revealed that the spatial and temporal limitations on movement, which lead to the detection of discontinuities, are absolute limits on each of the adjacent fields. That is, the movement of at least one of the two fields must be within the short range limits, in order to detect the boundary between them. If both fields undergo displacements larger than this limit, it is impossible to detect the discontinuity between them, regardless of their relative movement. Baker and Braddick (1982a) use this result to argue that the detection of discontinuities is not based on a computation that explicitly measures relative movement; rather, an absolute measurement of motion takes place first, followed by a process that compares nearby velocity measurements to locate the discontinuities. Van Doorn and Koenderink (1982, 1983) also reached this same conclusion, based on the results of experiments that assessed the ability to detect relative movement between two halves of a dynamic random dot pattern, to which spatiotemporal white noise is added.

Theoretically, a difference in speed alone is sufficient to define a boundary. Richards and Lieberman (1982) studied the detection of a motion boundary in random line patterns, for the case in which both fields move in the same direction, but with different speeds. The boundary was constructed both parallel to, and perpendicular to the common direction of motion. Their procedure tested the ability of the observer to localize a boundary, as a function of the percentage of moving elements in the display. Richards and Lieberman found that in general, a difference in speed is a poor cue to the location of boundaries; some observers were blind to boundaries created by shearing motion, in which the orientation of the boundary is parallel to the common image motion. Van Doorn and Koenderink (1982, 1983) found that when boundaries were defined by a change in speed alone, their detection depends primarily on the ratio of the difference in speed to absolute speed, and not on the absolute values of the two speeds. In particular, the difference in speed required for their discrimination task was roughly 60% of the average speed of the two adjacent fields.

In some further experiments with random dot patterns, Baker and Braddick (1982b) show that the ability to discriminate the orientation of a patch (that is, whether a rectangular patch is oriented horizontally or vertically), that moves against an uncorrelated background, varies very little with dot density, and increases with the size of the patch. The minimum patch size required for this discrimination task increases with the size of the discrete displacement of the patch. Baker and Braddick argued that this effect is a consequence of the fact that discrimination ability increases with eccentricity, due to increasing receptive field sizes. In related experiments by Chang and Julesz (1983), subjects were asked to discriminate the direction of movement of a patch of random dots that was embedded in a larger pattern of uncorrelated dots, as the area of the patch was varied. It was found that the maximum displacement that allowed correct direction discrimination increased with the area of the patch. Chang and Julesz also showed that the increase in receptive field sizes with eccentricity was by itself not sufficient to account for their experimental findings.

There are additional studies that address the ability to detect relative movement in random dot patterns (for example, Hadani, Gur and Meiri 1980; Nakayama and Tyler 1981; Nakayama 1981). We mentioned in the introduction that the ability to detect relative movement, and to use relative movement to localize boundaries, may be quite different. For example, if two adjacent fields are undergoing motion in the same direction, a 5% difference in speed is sufficient to detect relative movement (Nakayama 1981) (this result is similar to that of McKee (1981) for simple line stimuli). In the next section, the results of an experiment are presented, which suggest that the localization of a boundary may require a much larger speed difference, of roughly 50%.

## 7.2. The Localization of Boundaries in Random Dot Patterns

For many previous psychophysical experiments, it is difficult to assess whether the tasks involved address the ability to detect relative motion, or use relative motion to localize boundaries. In this section, we present an experimental technique that has been used to study the ability of the human visual system to localize the boundaries of discontinuous motion. We discuss some preliminary results of experiments that use this technique to explore some properties of the image and velocity field, on which the discontinuity detection depends. The author has served as the primary subject for these experiments, with other subjects observing the experimental setup on a more informal basis. For the results to be viable, further testing is required.

The first experiment, illustrated in Figure 7.1a, was designed to test the ability of observers to localize a boundary across which motion is discontinuous, as a function of the change in velocity across the boundary. The algorithms for discontinuity detection, discussed in Chapter 5, exhibit an increase in the ability to detect a discontinuity, as the change in direction of velocity between two adjacent surfaces increases. Algorithms that use measurements of the magnitude of velocity, as well as direction, will generally exhibit improved performance with increased changes in speed of motion across a boundary. In addition, for several types of detection algorithms, there are quantitative parameters, whose adjustment can alter the sensitivity of the algorithm to small changes in direction or magnitude of velocity. Thus, it would be useful to assess the ability of the human visual system to localize boundaries, as the change in velocity across the boundary is varied.

The stimuli in the experiment shown in Figure 7.1a were random dot patterns, whose dot size and density were varied. A sequence of frames was constructed in such a way that there was a discontinuity in the velocity field along a vertical line, whose horizontal position varied from one trial to the next. The position of the discontinuity was always within a distance of $2°$ of visual arc from the fixation point, either to the left or right. Along the boundary was a notch, whose size was $1° \times 2°$ of visual arc, and whose position was $1°$ above or below the point of fixation. The two fields of dots, to the left and right sides of the boundary, each underwent uniform translation, but with different velocities (they may have differed in direction or magnitude). The vertical boundary, with its notch, was not visible in any static frame; it was defined only by the difference in velocity to the left and

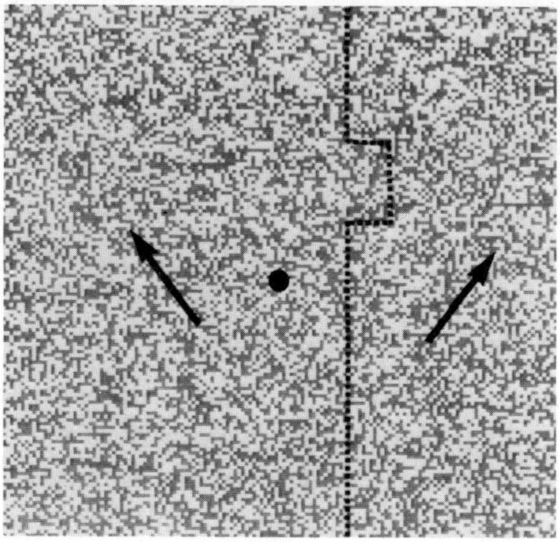

a.

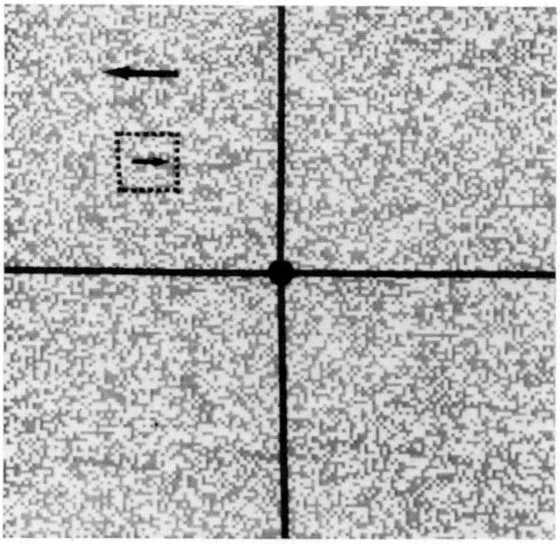

b.

**Figure 7.1. Experimental Stimuli.** (a) The differential movement of the two halves of the pattern reveals a boundary with a rectangular notch. The heavy dot indicates the fixation mark. (b) A small square patch of dots moves in a direction opposite the background. The heavy dot again indicates the point of fixation.

right of the boundary. The subject's task was to assess whether the notch on the boundary was above or below the point of fixation. This particular discrimination task requires scrutiny of the boundary; it cannot rely simply on an ability to detect a difference in the velocity between the two fields.

For the presentation of the stimulus, the subject maintained fixation at the center of the display (indicated by the heavy dot), and the first frame of a sequence appeared on the screen. On the subject's signal, the additional frames of the sequence appeared one after the other, in rapid succession. The interstimulus interval (ISI) was a few milliseconds (msec), and the total duration of the motion sequence was less than 100 msec, in order to remove the complicating effect of eye movements. To assess the ability of the subject to localize the boundary, a two-alternative forced-choice procedure was used. After each trial, the subject's task was to state whether the notch appeared above or below the fixation point. (Further details of the displays and experimental procedure are given in Appendix V.) For combinations of velocity that yield good discrimination, the boundary is quite strong, yielding nearly 100% correct judgements of the position of the notch. Two frames alone, with small ISI between, are sufficient to make this judgement. There are situations, however, in which relative movement between the two fields can be detected, but the boundary cannot be discriminated sufficiently well to determine the location of the notch.

We first explored the dependence of this boundary discrimination task on the difference in direction of velocity between the two fields. In these experiments, the individual dots subtended 3′ of visual arc, dot density was 50%, and viewing was monocular, from a distance of 30 cm. The entire random dot pattern subtended an area of size $8^\circ \times 8^\circ$ of visual arc. The size of the physical displacement of the dots varied from 3′ to 9.5′ of visual arc. The magnitude of the displacement was roughly equivalent for the two fields, but the difference in direction of motion varied over a small set of angles ($18.4^\circ, 26.6^\circ, 36.9^\circ, 45^\circ, 53.1^\circ, 61.4^\circ$). The small set of differences in direction is a consequence of moving the patterns along a discrete grid. The absolute directions of velocity, as well as relative directions, varied from one trial to the next (sampling the full 360° range), in order to prevent the subject from making anticipatory eye movements in the direction of an expected displacement (see Appendix V for details). Results from one subject (the author) appear in Figure 7.2. The percentage of correct responses is plotted as a function of

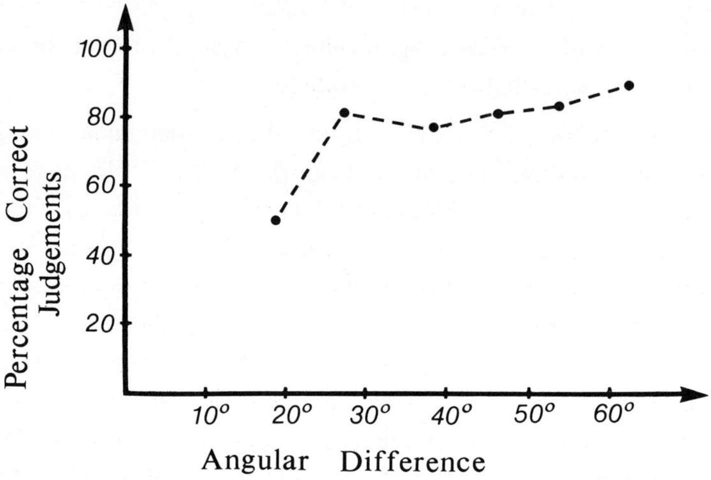


**Figure 7.2. The Dependence of Boundary Localization on Change in Direction.** The points indicate the percentage of correct judgements of the position of the notch as a function of the difference in direction of velocity across the boundary.

angular difference. Each data point represents the average of at least 50 trials. For the smallest angular difference, 18.43°, discrimination was at the level of chance. By the time the angular change reached 26.6°, performance was well above chance, and improved slightly as the angle increased further. These results suggest that the ability to localize a motion discontinuity depends on the change in direction of velocity across the boundary. For this subject, the minimum angular difference required for this discrimination lies somewhere between 18.43° and 26.6°.

Within the context of the dependence of boundary localization on the difference in direction of velocity, variations on the discrimination task were also considered. In one experiment, the width of the notch was varied between three discrete sizes, and the subject's task was to determine both the position and width of the notch. In a second experiment, the notch varied between a rectangular, triangular, and circular shape. The aim of both experiments was to determine the accuracy with which the boundary could be localized. The initial experimental data suggests that these more detailed discrimination tasks require a larger difference in direction of motion between the two fields.

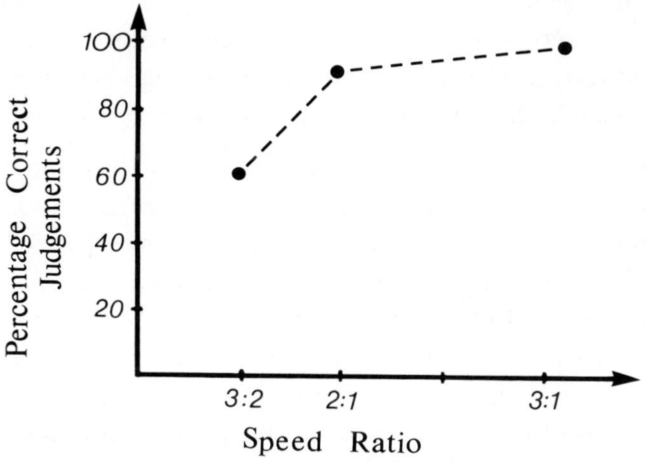


**Figure 7.3. The Dependence of Boundary Localization on Change in Speed.** The points indicate the percentage of correct judgements of the position of the notch as a function of the difference in magnitude of velocity across the boundary.

Previous experiments by Richards and Lieberman (1982) had shown that for some observers, a speed difference alone could define boundaries in the velocity field that the human visual system is capable of detecting. The dependence of boundary localization on speed differences was examined, with the same experimental procedure used previously. Figure 7.3 shows the percentage of correct judgements of the position of the notch, for speed ratios of 3:2, 2:1 and 3:1. Again, the author served as subject, and each data point represents the average of roughly 50 trials. While it is certainly possible to perform this discrimination on the basis of speed, the author required slightly better than a 3:2 speed ratio to reach the level at which judgements were 75% correct. This contrasts significantly with the 5% difference required to detect relative speed (Nakayama 1981).

Sparse random dot patterns, for example with a 10% density of dots, also yield the perception of sharp boundaries from discontinuities in the velocity field. Motion discontinuities in random dot patterns are often perceived as discontinuities in depth; the two adjacent fields are seen as opaque surfaces, with one surface occluding the other. Even in sparse dot patterns, the velocity field discontinuities are perceived as sharp discontinuities in depth. The position of the boundary in the image may vary, with its shape deviating from a straight line (depending on the position of the dots along the boundary), but the associated discontinuity in depth is always sharp, and very striking (similar to the boundaries perceived when viewing a sparse random dot stereogram (Julesz 1971)). Baker and Braddick (1982b) also found little variation in discontinuity discrimination with dot density.

To summarize this section, we have presented an experimental tool for studying the ability of the human visual system to utilize differential motion for localizing surface boundaries. Preliminary experimental data suggests that boundary discrimination depends both on changes in direction and magnitude of velocity across the discontinuity. Most discontinuity detection algorithms exhibit an increase in the ability to detect a discontinuity, as the change in direction of velocity of the two adjacent surfaces increases. It is certainly true of a scheme based on the initial measurements of $v^{\perp}(s)$ along zero-crossing contours, because the density of zero-crossing segments that yield a change in the sign of $v^{\perp}(s)$ across the boundary increases with the change in direction of velocity. For schemes that attempt to detect discontinuities after the computation of the velocity field, one would expect that the larger change in direction yields a more rapid variation in velocity direction across a surface boundary, which should be more easily detected. The fact that a difference in speed alone is sufficient to define a clear boundary implies that some measure involving the magnitude of velocity must be considered in the detection algorithm.

### 7.3. Velocity versus More Primitive Measurements

A key question for the algorithm utilized by the human visual system to detect motion discontinuities concerns the input representation for the detection process. Two possible representations are the initial measurements of the perpendicular components of velocity, and the full two-dimensional velocity field (other representations are also possible). If the detection of discontinuities begins to take place prior to the combination of the initial motion measurements for the computation of the velocity field, it might be possible to find a visual pattern for which discontinuities are easily detected, but for which the local velocity field cannot be made explicit. An example of such a pattern is shown in Figure 7.4. The direction of velocity changes across the pattern, in $45°$ increments, with the local direction of velocity of the top half always $180°$ out of phase with that of the bottom. This yields a discontinuity along the dashed line of Figure 7.4. The magnitude of displacement between frames was

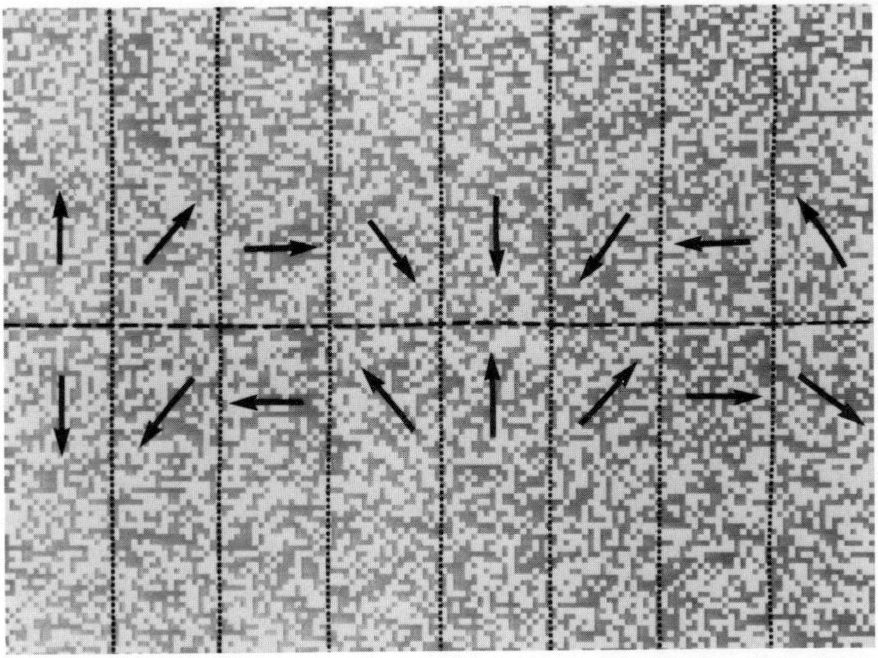

**Figure 7.4. A Complex Velocity Field.** The direction of velocity of rectangular patches is given by the arrows. The direction of velocity changes slowly from left to right across the pattern, but changes abruptly along the central horizontal line shown.

3′ of visual arc for the vertical and horizontal directions, and 4.24′ of visual arc for oblique directions. The dots subtended 3′ of visual arc, and each column of dots of a single direction had a width of 16 dots (48′ of visual arc). The full pattern subtended a square of size $16° \times 16°$ of visual arc in the center of the visual field.

When this pattern is observed in motion (with steady fixation at a point central to the display), either in a flash, or with prolonged viewing, the horizontal discontinuity across the entire pattern is very pronounced, but the directions of velocity of dots that are outside the immediate area of fixation are very difficult to judge. If a notch is placed on the edge, similar to the setup of Figure 7.1a, its position along the boundary can be judged quite easily. This result may support the idea that a precise velocity field computation need not precede the detection of motion discontinuities. In this example, the velocity field outside the immediate area of fixation is very difficult to describe, yet the boundary stands out clearly, extending several degrees from fixation.

### 7.4. Size versus Displacement

In Chapter 5, we discussed the limitations of a discontinuity detection scheme that is based on the initial measurements of motion along zero-crossing contours. Through a constraint on the density of zero-crossings, the size of the initial $\nabla^2 G$ operator determines the resolution at which the boundary may be localized, and the smallest object that can be discriminated against a textured background. In addition, the size limits the range of displacements that can be tolerated in discrete presentations, before the motion becomes incoherent. A trade-off results between the object size that can be discriminated, and the allowed displacements of the object between frames. This trade-off can be demonstrated psychophysically. Baker and Braddick (1982b) performed some experiments in which the subject was asked to discriminate the orientation of a patch of random dots that moves coherently against a background of uncorrelated dots. They showed that the minimum patch size required for this discrimination task increased with the size of the displacement. We experimented with the object size that can be discriminated against a background, in a dense random dot pattern, using a different experimental setup.

The stimulus for this experiment is shown in Figure 7.1b. As before, the subject maintained fixation on a point at the center of the display, and a sequence of frames was presented in a flash. In this case, the entire field moved in one of four directions (right, left, up or down), and a small patch of dots appeared in one of the four quadrants, moving in the opposite direction. The magnitude of displacement of the patch and surround was always equal. The patch was positioned a distance of $1^\circ - 2^\circ$ of visual arc from the fixation point. Its size, displacement between frames, and position varied between trials. Each dot subtended $3'$ of visual arc, and displacement of the patch between frames varied from one to four dots. (Further details are presented in Appendix V.) Experimental results, with the author as subject, are displayed in Figure 7.5. Each data point represents the results of at least 50 trials. The percentage of correct judgements is plotted against the size of the patch, for four displacements: $3'$ (filled circles), $6'$ (crosses), $9'$ (open circles), and $12'$ (stars) of visual arc. Two observations can be made. First, the patch size that can be discriminated is very small. For a $3'$ displacement, an $18' \times 18'$ patch was localized

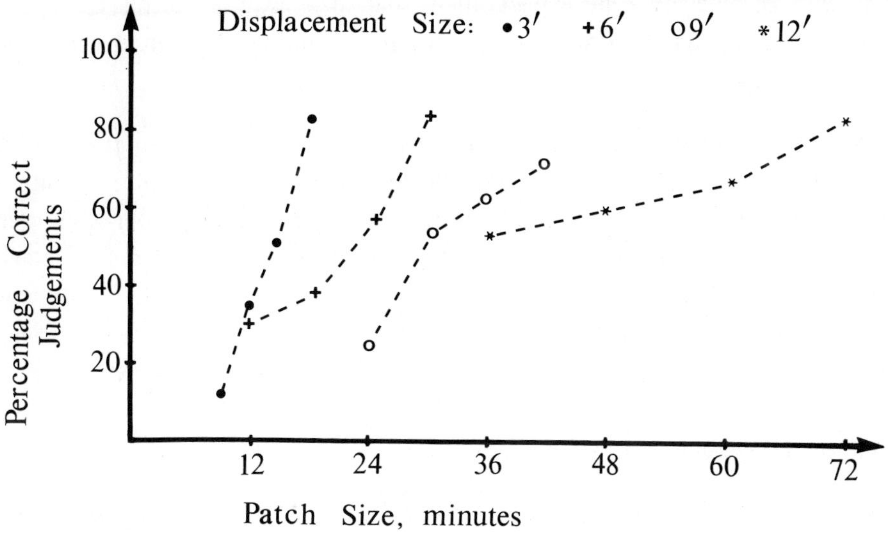


**Figure 7.5. Size versus Displacement.** The points indicate the percentage of correct judgements of the position of the patch as a function of patch size, for four different displacement sizes.

correctly in 80% of the trials. Second, the patch size required for discrimination of position increases with the size of the displacement. For a particular level of correct responses (such as 75%), the patch size required to achieve this level increases with displacement.

This result is expected for a system that utilizes a range of operator sizes in its analysis of motion. As displacement increases, the smaller operators begin to measure incoherent motion. As the smallest size operator that can detect coherent motion increases (due to increasing displacement), the density of zero-crossing measurements on which the detection could be based decreases, so that a larger object is required to make a reliable judgement of the discontinuity. In addition, the inconsistency of information from smaller operators could contribute to the greater difficulty in discriminating the pattern at larger displacements. These results are consistent with those of Baker and Braddick (1982b) and Chang and Julesz (1983). As Baker and Braddick suggest, the increase in the discrimination ability with eccentricity may also contribute to an improved detection with patch size. The early operators underlying the spatial frequency channels in human vision increase in size with eccentricity. One expects then, that the maximum displacement size required for the perception of coherent motion in random dot patterns increases with eccentricity.

With regard to the computations underlying the detection of motion discontinuities, these experiments also suggest that this detection may involve operations of limited spatial extent. For example, if smoothing or averaging operations were used, as discussed in Chapter 5, the spatial extent of the operations cannot be very large, compared with the size of the initial $\nabla^2 G$ operator. The results support the idea that multiple size operators are utilized in this analysis, but do not yet provide insight into the way in which the results from different operator sizes are integrated.

### 7.5. Future Psychophysical Experimentation

We mentioned in Section 5.3 that one limitation of a scheme that relies on the initial motion measurements along zero-crossing contours arises from the confounding of orientation and velocity of the zero-crossings. In general, as the change in direction of velocity across the discontinuity decreases, the density of zero-crossings that could be used to detect the boundary also decreases. If, in addition, the image structure is such that there exists relatively few zero-crossings at the appropriate orientations, detection is difficult. A possible area for psychophysical investigation is the relationship between detection ability, and the distribution of the orientation of intensity changes in the image. There are at least two methods for manipulating the distribution of zero-crossing orientations. First, a pattern of random line segments could be used, rather than a pattern of dots. The experimental question is whether the ability to detect a discontinuity depends on the orientation of the lines, relative to the direction of velocity of the two adjacent fields. A second method is to filter a random dot pattern with an oriented smoothing operator (such as an oriented Gaussian filter), and test whether detection depends on the orientation of the filter. In both cases, if there is no change in detection ability, the results are inconclusive, because there do exist alternative sources of information for the detection process (such as the line endpoints, for patterns consisting of oriented line segments). If there does exist a change in the ability to detect a discontinuity, it may support an algorithm based on the initial motion measurements.

A second issue for future exploration is the accuracy with which the boundary is localized. For example, using the basic setup of Figure 7.1a, the size and shape of the notch could be varied, and the subject required to make a more detailed assessment of the boundary. One could also explore random dot patterns in which the border has a more complex shape, such as a sine wave. Preliminary experiments suggest that even in a flash, the visual system appears to have little difficulty in perceiving more complicated boundaries. The precise localization of a boundary could form the basis for a comparison between the results of a possible algorithm and human performance.

Finally, adjacent surfaces may rotate as well as translate. In the case of a rotating circle against a translating background, two frames in a brief flash are sufficient to define a clear boundary. For the varying velocity field of Figure 7.5, while the velocity field was difficult to describe, the boundary was extracted quite easily. In Chapter 5, it was shown that rotations complicate the detection process. The ability of the human visual system to easily detect boundaries between rotating surfaces presents a challenge for the development of an algorithm that exhibits similar competence to that of the human system.

# 8. NEUROPHYSIOLOGICAL STUDIES OF MOTION

With regard to the analysis of visual motion, four specific computations have been discussed in this book:

(1) The convolution of the image with the $\nabla^2 G$ operator, and measurement of its temporal derivative.

(2) The measurement of $v^{\perp}(s)$ along zero-crossing contours.

(3) The combination of the measurements of $v^{\perp}(s)$ along contours, to compute the two-dimensional velocity field.

(4) The detection of motion discontinuities.

In this chapter, we discuss some relevant neurophysiological studies of visual areas that may be involved in these computations: the retina and lateral geniculate nucleus (LGN), striate visual cortex, and the middle temporal area (MT) of extrastriate cortex, located in the posterior bank of the superior temporal sulcus (STS). For each visual area, we first review some relevant anatomical and neurophysiological studies, and then discuss its possible role in the above computations.

The motivation of this work has been to contribute toward an understanding of the way in which movement is analyzed by the human visual system. The results of neurophysiological studies, particularly those of the primate visual system, aid a computational study by suggesting the possible algorithms that the system uses in performing a given computation. For example, the problem of edge detection was studied extensively in the field of computer vision, and gave rise to several methods for the automatic detection of intensity changes. Many aspects of the schemes proposed by Marr and Poggio (1979) and Marr and Hildreth (1980) had been studied previously (for a review, see Hildreth (1983)). Physiological studies of the retina and visual cortex, discussed in Sections 8.1 and 8.2, suggested the particular set of operations that may be used in higher biological vision systems. These physiological studies, together with psychophysical studies of early visual processing, then led Marr, Hildreth and Poggio to examine the implication of these operations for the detection of intensity changes.

Neurophysiological investigations may also benefit from computational studies, in at least two ways. First, a computational study may help to refine a model of the function of a particular class of neurons. Second, by elucidating the specific computations that are necessary for general visual processing, these studies may help to guide an initial neurophysiological exploration of possible computations performed by a visual area. As an example, it has been suggested since the early studies of Hubel and Wiesel (1962, 1968) that the class of *simple cells* in visual cortex may be involved in the detection of edges, or features, in the image. On the basis of computational studies, it was proposed that at least one class of simple cells may be specifically involved in the detection of zero-crossings in the output of the $\nabla^2 G$ convolution. This model served as motivation for a recent neurophysiological study, described in Section 8.2, which supports this view (Richter 1982).

### 8.1. The Retina– $\nabla^2 G * I$ and its Time Derivative

In the human visual system, when light first reaches the eye, it is immediately processed by a number of interconnected layers of cells in the retina. The result of this processing is represented in the output of the retinal ganglion cells, whose fibers form the optic nerve, which carries visual information to the lateral geniculate nucleus (LGN), from which it is relayed to visual cortex. Early electrophysiological studies of retinal ganglion cells in the cat showed their receptive fields to have an antagonistic center-surround organization (Kuffler 1953). Rodieck and Stone (1965) suggested that this organization was the result of superimposing a small central excitatory region on a larger inhibitory 'dome' that extends over the entire receptive field. Rodieck (1965) and Enroth-Cugell and Robson (1966) described the two 'domes' as Gaussians, thus describing the overall shape of the receptive field as the difference of two Gaussians (DOG). In analytic terms, Rodieck (1965) described the output of the retinal ganglion cells as the convolution of the changing image with the spatial DOG operator, combined with a particular temporal filtering.

The retinal ganglion cells do not form a single homogeneous class; two of the main cell types have been labelled X and Y cells (Enroth-Cugell and Robson 1966; Cleland, Dubin and Levick 1971). Some of the physiological distinctions between the two classes are the following:

(1) **Spatial Summation:** X cells exhibit linear summation of their inputs, while Y cells show nonlinear summation (*cat:* Enroth-Cugell and Robson 1966; Hochstein and Shapley 1976; *monkey:* Schiller and Malpeli 1977; DeMonasterio 1978b).

(2) **Temporal Response to Light Stimulation:** X cells show a sustained response to changes in light intensity, while Y cells exhibit a transient response (*cat:* Cleland, Dubin and Levick 1971; Cleland, Levick and Sanderson 1973; *monkey:* DeMonasterio 1978b).

(3) **Receptive Field Size:** Y cells have larger receptive fields than X cells (*cat:* Enroth-Cugell and Robson 1966; Boycott and Wassle 1974; Peichl and Wassle 1979; *monkey:* DeMonasterio and Gouras 1975).

(4) **Color Specificity:** The center and surround responses are cone specific in X cells, with the cone types often spectrally different, while the spectral sensitivity of Y cells is generally broad band (*monkey:* DeMonasterio and Gouras 1975; Schiller and Malpeli 1977; DeMonasterio 1978a,b).

(5) **Conduction Velocity:** Y cells exhibit a faster axonal conduction than X cells (*cat:* Cleland, Dubin and Levick 1971; Dreher, Fukada and Rodieck 1976; *monkey:* Sherman, Wilson, Kaas and Webb 1976; Schiller and Malpeli 1977; DeMonasterio 1978a).

X and Y cells are also grouped into ON-center and OFF-center cells, depending on whether they are excited by light increment or decrement in the center of their receptive fields.

In addition to the X and Y cells, there are other cell types in the retina that have been variously labelled as W cells (Stone and Hoffman 1972), *rarely encountered* cells (Cleland and Levick 1974; Schiller and Malpeli 1977), and Type II, V and VI cells (DeMonasterio 1978c). This group of cells generally exhibits a slower axonal conduction velocity than X and Y cells (Stone and Hoffman 1972; Cleland and Levick 1974). At least two types of receptive field properties have been observed. One group of cells exhibits considerable spontaneous activity, and is suppressed by intensity edges in their receptive field (Rodieck 1967; Stone and Hoffman 1972). A second group exhibits low spontaneous activity, and is excited by spots of light of any contrast, flashed anywhere in their receptive field (Stone and Fabian 1966; Stone and Hoffman 1972). Other cell types have been observed by Cleland and Levick (1974) in the cat, and DeMonasterio (1978c) in the monkey.

Anatomically, X and Y cells have different dendritic morphologies, and project to different areas in the LGN and cortex (*cat:* Boycott and Wassle 1974; Cleland and Levick 1974; Fukada and Stone 1974; Wassle, Levick and Cleland 1975;

*monkey:* Hubel and Wiesel 1972; DeMonasterio and Gouras 1975; Dreher, Fukada and Rodieck 1976; Sherman, Wilson, Kaas and Webb 1976; Schiller and Malpeli 1977, 1978; DeMonasterio 1978a,c; Rodieck 1979). The class of W cells also differs anatomically from the X and Y cells (Boycott and Wassle 1974), and may project primarily to the superior colliculus (Schiller and Malpeli 1977; DeMonasterio 1978c).

At the LGN, the main properties of the X and Y cells are preserved (Cleland, Dubin and Levick 1971; Hoffman, Stone and Sherman 1972; Dreher and Sanderson 1973). One difference between retinal and geniculate cell responses is a lower maintained discharge to uniform illumination in LGN cells (Hubel and Wiesel 1961; Cleland, Dubin and Levick 1971). This may be the result of a limited convergence of retinal ganglion cells on geniculate cells; it has been shown that geniculate cells are driven by only one, or a very few, retinal ganglion cells (Singer and Creutzfeldt 1970; Levick, Cleland and Dubin 1972). This limited convergence may also result in a broader spread of receptive field sizes in the LGN (Marr and Ullman 1981).

Marr and Hildreth (1980) proposed that the output of the retinal ganglion X cells represents the convolution of the image with the $\nabla^2 G$ operator, for the purpose of detecting intensity changes. The $\nabla^2 G$ operator may be implemented in neural hardware as the difference of two Gaussians; the DOG provides a close approximation to $\nabla^2 G$, for a particular choice of space constants (Appendix B, Marr and Hildreth 1980). The positive and negative components of this convolution are assumed to be carried by the ON-center and OFF-center cells. Marr and Ullman (1981) proposed that the output of the Y cells represents the temporal derivative of the $\nabla^2 G$ convolution, which may be useful for the detection of motion.

These models of the function of X and Y cells led Richter and Ullman (1980, 1982) to propose a detailed neural model for the derivation of their spatial and temporal properties, via interactions between cell layers in the retina. The X and Y cell responses predicted by their model agree well with available electrophysiological recordings, for a variety of stimuli, providing support for the computational functions proposed by Marr and Hildreth (1980) and Marr and Ullman (1981). It should be noted that the computations of the $\nabla^2 G$ convolution and its temporal derivative have been proposed as primary functions for retinal cells, but they are certainly not their only function. A potential role in these two computations does not exclude an additional role of these cells in computations such as brightness normalization, the reduction of noise sensitivity, or the analysis of color.

### 8.2. Visual Cortex– The Measurement of $v^{\perp}(s)$ Along Zero-crossings

The output of the LGN forms the primary upward-flowing input to visual cortex. If the X and Y cell outputs represents the convolution of the image with $\nabla^2 G$ (or the DOG approximation) and its temporal derivative, and these outputs are essentially preserved in the LGN, then these two representations of the changing image would form the primary input to visual cortex. It has been proposed that the first step in the analysis of motion is the measurement of $v^{\perp}(s)$ along zero-crossing contours in the $\nabla^2 G$ convolution output. This computation involves the detection of zero-crossings, and measurement of their movement over time. In this section, we briefly discuss evidence for the implementation of this computation by a class of cells in primary visual cortex.

In early studies of cortical cells in the cat and monkey, Hubel and Wiesel (1962, 1968) isolated four main functional classes, which they labelled as nonoriented, simple, complex and hypercomplex cells. Nonoriented cells, found mainly in layer IV (the input layer to visual cortex), have receptive field properties similar to those of incoming geniculate fibers. A simple cell generally responds optimally to an edge or bar of a particular orientation and sign of contrast, moving across its receptive field. Many simple cells are also selective for the direction of motion of a stimulus. Complex cells show similar specificities. The main distinction between the two classes is that simple cells have spatially distinct regions within their receptive field, for which either light increment or decrement (but not both) elicit a response, while complex cells respond both to light increment and decrement throughout their receptive fields. The criterion that distinguishes hypercomplex cells is an inhibition of the cell's response when the length of a stimulus is extended along the axis of orientation of the cell, rendering it selective for stimulus length.

In a quantitative study of cortical cells in the rhesus monkey, Schiller, Finlay and Volman (1976a) further subdivided the simple and complex cell classes, on the basis of the spatiotemporal distribution of their response to moving edges and rectangles, and stationary flashed stimuli. The two main classes were labelled S-type and CX-type cells; their study supports the property used by Hubel and Wiesel as the primary distinguishing factor between the two classes. Within the S-type category, however, Schiller, Finlay and Volman further identified 7 distinct cell types (Bishop, Coombs and Henry (1971a,b) show a similar variety of cell types in the visual cortex of the cat). Of interest here are the classes labeled $S_1$ and $S_3$. $S_1$ cells respond to a single edge of a particular orientation and sign of contrast, moving in a preferred direction. The class of $S_3$ cells are selective for an edge of a particular contrast sign, moving in either direction. Other classes of cells within the S-type category exhibit a more complicated spatiotemporal distribution in their response to simple stimuli. Poggio, Doty and Talbot (1977) draw an analogy between the S-type cells described by Schiller *et al.*, and a class of cells they refer to as *modulated* cells, whose response modulates with the temporal frequency of moving gratings of black and white bars. These modulated cells are highly responsive to moving gratings, and in general, less responsive or insensitive to stationary patterns.

The two classes of S-type cells noted above exhibit properties that are expected of cells whose function may be, at least in part, the detection of oriented zero-crossing contour segments. The directional selectivity of $S_1$ cells suggest a possible role in the measurement of zero-crossing motion. Movshon *et al.* (1983) have shown that the nature of the directional selectivity in cortical neurons of the cat and monkey is such that these neurons can only signal the component of motion in the direction perpendicular to the orientation of spatial features in the image. In a recent study of cortical cells in the cat, Burr, Morrone and Maffei (1981) show that simple cells are unresponsive to a moving two-dimensional textured pattern, yet respond well to a one-dimensional pattern which, in a Fourier sense, is contained within the two-dimensional pattern. They provide evidence that this behavior of simple cells is the result of inhibition from cells tuned to orientations outside the selectivity band of the simple cell. DeValois, Yund and Hepler (1982) show similar inhibition in cells of the macaque visual cortex. Burr, Morrone and Maffei argue that this inhibition could serve to improve the selectivity of simple cells to visual contours, by suppressing responses to textured patterns.

The potential role of simple cells in the detection of zero-crossings was examined in a recent neurophysiological study by Richter (1982; Richter and Ullman, in preparation). The basic two-dimensional stimulus, illustrated in Figure 8.1a, consists of two adjacent step changes of intensity at a particular orientation. In Figure 8.1b, the one-dimensional cross-section of the stimulus is shown, for three different distances between the two edges. The convolutions of the profiles in Figure 8.1b with

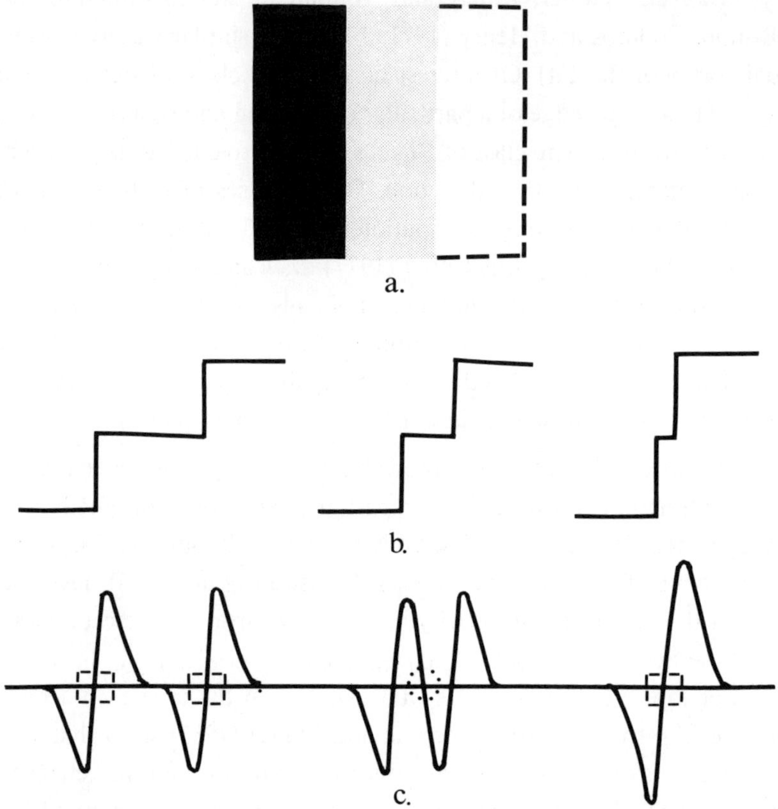


**Figure 8.1. The Zero-crossing Experiment.** (a) The visual stimulus, consisting of two adjacent edges. (b) The one-dimensional cross-sections of the stimulus, for three different separations between the edges. (c) The convolutions of the profiles of (b) with a $D^2G$ operator.

a particular size $D^2G$ operator, are shown in Figure 8.1c. When the distance between the two edges is large (leftmost profile), there are two distinct zero-crossings in the convolution output, as indicated by the dotted squares. When the edges are very close (rightmost profile), a single zero-crossing results. There is an intermediate distance (middle profile) that yields three zero-crossings, with the center zero-crossing, indicated by the dotted circle, of the opposite sign of contrast. Suppose that a particular directionally selective simple cell responds only to movement to the right, of an edge that changes from dark to light as we move from left to right in the image. Richter (1982) found that such a cell would respond to the double-edge stimulus, at the locations of the zero-crossings indicated by the dotted squares, when the three patterns were moved to the right. Now, suppose that the sign of contrast of the two edges is reversed, and the resulting stimulus is moved to the right. For the intermediate distance between the two edges, the middle, spurious zero-crossing is now of the appropriate sign of contrast for the cell. Richter found that in the visual cortex of the cat, many simple cells responded to this reversed pattern, at the location of this additional zero-crossing. This phenomenon supports the role of simple cells in the detection of zero-crossings; in the reversed contrast stimulus, there are no edges of appropriate contrast sign for the cell, but due to the particular spatial arrangement of the two edges, there does exist a zero-crossing of the appropriate sign of contrast to trigger the cell's response.

If part of the function of $S_1$ cells were to measure the perpendicular component of velocity along zero-crossing contours, one can also examine the extent to which these units could represent the magnitude of $v^{\perp}(s)$. Experimental studies indicate that the magnitude of the simple cell response exhibits a somewhat complex dependence on the spatial and temporal frequency, orientation, size and contrast of a visual pattern (for example, Tolhurst and Movshon 1975; Schiller, Finlay and Volman 1976a,b; Tolhurst, Movshon and Thompson 1981; Albrecht and Hamilton 1982; DeValois, Albrecht and Thorell 1982; DeValois, Yund and Hepler 1982). With regard to their sensitivity to speed, Orban, Kennedy and Maes (1981) found four types of speed tuning in cortical cells of the cat; cells responded only to low speeds (low-pass), only to high speeds (high-pass), were selective to a particular narrow range of speeds, or were very broadly tuned in their speed sensitivity. The optimal speeds for the narrowly tuned cells ranged from $3° - 120°$/second in the central part of the visual field; in the periphery, cells tuned to low speeds were absent. It was also found that the width of speed tuning decreased with increasing optimal speed. These studies suggest that there may at least be a rough measurement of the magnitude of $v^{\perp}(s)$ available for later processing.

There are many possible mechanisms by which the detection of moving zero-crossings could be implemented in neural hardware. Two such mechanisms, proposed by Marr and Ullman (1981), and Poggio (1983), were discussed in Section 6.1.3, and illustrated in Figure 6.10. These models are as yet too simple to account for the range of physiological data available on the behavior of simple cells, but this does not rule out a possible role in zero-crossing detection; it only suggests that the particular mechanism involved may be quite complex.

## 8.3. Higher Cortical Areas– The Velocity Field and its Discontinuities

The combined orientation and direction selectivities of units in striate cortex suggest that they are only capable of measuring the component of motion in the direction perpendicular to the orientation of spatial features in the image. We have argued that a second stage of analysis is required, which combines the responses of these directionally-selective units to compute a two-dimensional velocity field, that is independent of image structure. Beyond striate cortex, there is a small visual area in the posterior bank of the superior temporal sulcus (STS) of the macaque, referred to as area MT (Gattas and Gross 1981; Van Essen, Maunsell and Bixby 1981a,b; Weller and Kaas 1981), whose physiological properties suggest a role in this combination stage (Zeki 1974, 1978; Van Essen, Maunsell and Bixby 1981a; Albright and Gross 1982; Newsome, Mikami and Wurtz 1982, Maunsell and Van Essen 1983). Three main properties of MT make it suitable for motion analysis: (1) a direct, topographic projection from striate cortex, (2) a large proportion of cells that are selective for the direction of motion of a stimulus, with little selectivity for its spatial structure, and (3) efferent projections to other areas, both cortical and subcortical, which play a role in motor functions.

Similarities between this area of the macaque visual cortex, and area MT of other primates has suggested a homology (for discussion, see Van Essen, Maunsell and Bixby 1981a,b). Anatomically, both areas are distinguished by heavily myelinated fibers, and a direct projection from striate cortex. There is some controversy over the precise extent of MT in the macaque. Some researchers maintain that the myelinated area within STS exactly coincides with the striate projection zone, and defines the area MT (Van Essen, Maunsell and Bixby 1981a). Others maintain that the striate projection zone within STS extends slightly beyond the myelinated region, in the medial direction, and include the entire projection zone in their definition of area MT (Ungerleider and Mishkin 1979; Gattas and Gross 1981). In the owl monkey, area MT is more clearly defined; physiological studies there have also suggested a role in the analysis of visual motion (Zeki 1980; Baker, Petersen, Newsome and Allman 1981; Miezin, McGuiness and Allman 1982).

### 8.3.1. The Striate Projection to Area MT

Area MT receives a direct projection from striate cortex. Anatomical studies indicate that this projection arises from stellate and pyramidal cells in layer IVb, and pyramidal cells in layer VI (*macaque:* Lund *et al.* 1975; Lund *et al.* 1979; *marmoset:* Spatz 1977). A study by Dow (1974) suggests that there may be a large proportion of directionally selective cells in layer IVb. It is useful to consider the possible source of inputs for cells in this layer. Dreher, Fukada and Rodieck (1976) and Sherman *et al.* (1976) show a segregation of X-type and Y-type cells in the LGN, into the parvocellular and magnocellular layers, respectively. The parvocellular and magnocellular layers then project separately to striate cortex, terminating in layers IVc$\beta$ and IVc$\alpha$, respectively (Hubel and Wiesel 1972; Lund and Boothe 1975; Lund *et al.* 1975). Within striate cortex, layer IVc$\alpha$ projects heavily to layer IVb, while IVc$\beta$ projects heavily to layer III (see, for example, Lund *et al.* 1979). Both stellate and pyramidal cells appear to project to MT (although in the squirrel monkey, most of the projection cells are pyramidal (Tigges *et al.* 1981)). Stellate cells receive their input primarily from within the single layer IVb (Lund *et al.* 1975), suggesting that the source of their input may be primarily Y-type units. The pyramidal cells, however, receive connections from layer IVb, as well as the superficial layers I, II and III (Lund *et al.* 1975; Lund *et al.* 1979). These pyramidal cells may therefore have access to both X-type and Y-type inputs, and may be possible candidates for the implementation of the Marr-Ullman simple cell model. The pyramidal cells of layer VI, which also project to MT, receive their inputs primarily from the deeper layers V and VI, which contain terminals from X-type and Y-type input fibers, stellate cells from layers IVc$\alpha$ and IVc$\beta$, and cells in superficial striate layers (Lund *et al.* 1979).

The mapping from striate cortex (area 17) to MT is topographic (Zeki 1974, 1978; Weller and Kaas 1978; Gattas and Gross 1979, 1981; Ungerleider and Mishkin 1979; Montero 1980; Van Essen, Maunsell and Bixby 1981a,b); connections are made between units whose receptive fields are located in roughly the same location in the visual field. Thus, MT contains a retinotopic representation of visual information. Like area 17, a much larger portion of MT is devoted to a representation of the central visual field (Gattas and Gross 1979, 1981). Van Essen, Maunsell and Bixby (1981a) estimate that the representation of the central 15° of the visual field occupies slightly more than half of MT. The representation in MT, however, is coarser than that of area 17 (Zeki 1974). At any location in the visual field, the receptive fields of cells in MT are larger than those of corresponding cells in area 17, and the sizes of receptive fields in MT increase faster with eccentricity (Gattas and Gross 1981). The far peripheral visual field is included in large receptive fields whose centers are located at eccentricities of 30° to 50°. There also appears to be a denser representation of the lower half of the visual field in MT (Gattas and Gross 1981; Van Essen, Maunsell and Bixby 1981a). Van Essen *et al.* found only 26% of the units they studied to have receptive fields above the horizontal meridian. The results of anatomical and physiological studies, taken together, indicate that MT contains a complete, retinotopic representation of information in the contralateral visual field, whose resolution decreases with eccentricity, similar to striate cortex.

### 8.3.2. Other Connections of Area MT

In addition to the striate projection, MT in the macaque also receives input from area V2 (area 18) (*macaque:* Maunsell and Van Essen 1982; Ungerleider, Desimone and Mishkin 1982; *other primates:* Tigges, Spatz and Tigges 1974; Martinez-Millan and Hollander 1975; Tigges *et al.* 1981), areas V3, VP, and a region anterior to VP in ventral extrastriate cortex (Maunsell and Van Essen 1982), and the inferior and lateral subdivisions of the pulvinar (Benevento and Rezak 1976; Van Essen, Maunsell and Bixby 1981a; Maunsell and Van Essen 1982; Standage and Benevento 1982). In the macaque, cells in the lateral pulvinar have physiological properties that suggest a role in the control of saccadic eye movements (Petersen, Robinson and Keys 1982).

Efferent connections from MT reach both cortical and subcortical areas. There is a small reciprocal projection to striate cortex and area V2 (*macaque:* Weller, Graham and Kaas 1979; Van Essen, Maunsell and Bixby 1981a,b; Maunsell and Van Essen 1982; Underleider, Desimone and Mishkin 1982; *other primates:* Tigges *et al.* 1981; Weller and Kaas 1981). The major cortical efferents of MT project to other areas within STS, and the intraparietal sulcus. Maunsell and Van Essen (1982) describe a single projection area medial to MT in STS, which they refer to as the medial superior temporal area (MST). There is no clear topographic organization of the projection from MT to MST. Ungerleider, Desimone and Mishkin (1982) distinguish two projection areas within STS. The first is located in the posterior bank of STS, bordering the lower field representation of MT, and receives a projection only from MT sites representing the lower visual field and center of gaze. The second area surrounds MT anteriorly and medially, and receives a highly convergent input from MT with only a "suggestion" of topography. The relationship between the single area observed by Maunsell and Van Essen, and the two areas observed by Ungerleider *et al.* is not clear. Maunsell and Van Essen also describe a single projection area within the intraparietal sulcus, which they refer to as the ventral intraparietal area (VIP). There is again no obvious topographical organization of the projection from MT to VIP. Ungerleider, Desimone and Mishkin (1982) distinguish two areas within the intraparietal sulcus, which both receive convergent input from all parts of MT. Finally, Maunsell and Van Essen (1982) observed connections between MT and area V4, which were not clearly forward or backward connections.

Subcortical areas that receive a projection from MT (*macaque:* Maunsell and Van Essen 1982; *owl monkey:* Weller and Kaas 1981) include the inferior and lateral subdivisions of the pulvinar (Standage and Benevento 1982), pontine nuclei (which project to the cerebellum), pregeniculate nucleus, claustrum, reticular nucleus of the thalamus, superior colliculus, and the caudate nucleus and putamen within the basal ganglia. We have already noted the possible role of the pulvinar in the control of eye movements (Petersen, Robinson and Keys 1982). Many of the other subcortical areas receiving a projection from MT are also involved in various aspects of motor control (for example, see Mountcastle 1974, Part V).

### 8.3.3. Physiological Properties of MT Neurons

The main physiological properties that have been studied for cells in MT of the macaque are direction and orientation selectivity, using spots and bars of light of various sizes, moving across the visual field. One difference between MT units and directionally selective units in area 17, that has been revealed by these studies, is a selectivity for the direction of motion of a stimulus, with less selectivity for its spatial structure. Zeki (1974, 1978) first observed this behavior in MT units of the rhesus monkey. Using rectangles and bars of different dimensions, Zeki found that many MT units responded optimally to movement in a particular direction, regardless of the orientation and size of the visual stimulus. Later studies suggested similar behavior in MT neurons (*macaque:* Gattas and Gross 1981; Van Essen, Maunsell and Bixby 1981a; Albright and Gross 1982, Maunsell and Van Essen 1983; *other primates:* Zeki 1980; Baker, Petersen, Newsome and Allman 1981).

Van Essen, Maunsell and Bixby (1981a) make a distinction between directionally selective, directionally biased, and nonoriented cells. Directionally selective cells are those for which there is one direction of motion which yields a maximum response, with the opposite direction of motion yielding no response. Directionally biased are those that exhibit a larger response for movement in the optimal direction, and nonoriented cells show no preference for direction of motion. Of the 181 cells they studied, 79% were directionally selective, 18% directionally biased, and only 3% were nonoriented. The stimuli in these studies were oriented bars, which were always moved in the direction perpendicular to their orientation. These statistics compare well with those of Albright and Gross (1982) and Zeki (1978) for directionally selective units in MT.

MT units are not uniform in their selectivity for direction and orientation. Albright and Gross (1982) measured direction tuning curves, using both moving spots and bars as stimuli. In the case of moving spots, they found the mean width of direction tuning curves (the width of the tuning curve at half the maximum response of the cell) to be 108°, with a range of bandwidths from 22° to 279°. The statistics were similar for moving bars; mean bandwidth was 85°, with a range from 29° to 261°. Some units were very narrowly tuned for direction of motion, while others were very broadly tuned. Maunsell and Van Essen (1983) also observed a wide range in the direction tuning of MT units, using moving bars as stimuli. Baker *et al.* (1981) observed fairly broad direction tuning for MT units in the owl monkey. The orientation tuning of MT units has been measured with stationary flashed bars. Albright and Gross (1982) found a greater selectivity for orientation with this type of stimulus. The mean bandwidth for orientation tuning curves was 49°, with a range from 9° to 133°. Some units exhibited both direction and orientation selectivities, similar to cells in striate cortex. Van Essen, Maunsell and Bixby (1981a, see also Maunsell and Van Essen 1983) and Baker *et al.* (1981) also found clear orientation selectivity for many MT units, when using stationary flashed bars as stimuli.

The optimal speed of movement also varies between units. Baker *et al.* (1981) found that MT units in the owl monkey generally responded best to speeds ranging from 5° — 50°/second; a similar range of optimal speeds was found for MT units in the rhesus monkey (Zeki 1978). Some units in the owl monkey responded optimally to speeds larger than 100°/second, with one unit responding to a speed of 500°/second. Maunsell and Van Essen (1983) found that roughly 80% of the units they studied in MT responded well only to a limited range of speeds. Some units were fairly narrowly tuned for speed, and many showed inhibition to speeds that were far from their preferred speed. The overall distribution of preferred speeds spanned a wide range, from 2° — 256°/second, with a peak near 32°/second.

In a more recent study, Movshon *et al.* (1983) compared the responses of MT units to two different visual stimuli. The first consisted of a single oriented sine wave grating moving across the visual field, and the second was a combined pattern of two sine wave gratings of different orientations, each moving in the direction perpendicular to their orientation. The combined pattern was similar to that illustrated in Figure 6.20a. This comparison revealed two classes of neurons in MT. The responses of the first class suggest that they represent only the component of motion in the direction perpendicular to a grating of a particular orientation. The second class, however, appears to represent the two-dimensional direction of motion of the combined pattern. This second class may therefore be involved in the combination of local motion measurements to compute the two-dimensional velocity field.

Newsome, Mikami and Wurtz (1982) studied the response of MT units in the macaque to apparent movement stimuli. A square was moved across the visual field in discrete steps. With very low presentation rates, for which the interflash interval, IFI, was less than 150 msec, MT units responded transiently to each flash, regardless of the direction of motion of the square. As the IFI was increased, a transition point was reached when the unit became directionally selective, responding only to its preferred direction of motion. For appropriate spatial displacements between presentations, the directionally selective response could be elicited for IFIs up to 70 msec (occasionally up to 200 msec). For appropriate IFIs, directionally selective responses were generally obtained for displacements up to 1.5° of visual arc. The IFIs are commensurate with the temporal limitations of the short range process in human vision, measured psychophysically. The displacement limits may be related to the size of the underlying directionally selective units, also reflecting the limits of the short range process. The psychophysically measured spatial limit only applies to the central fovea, so cannot be compared with the displacement limit observed in this study, which may have included MT units whose receptive fields were in the periphery, where the directionally selective units are larger.

Some studies of the physiological properties of MT neurons in the owl monkey have used random dot patterns as stimuli. Baker *et al.* (1981) found that for some cells, this was a more effective stimulus than moving spots and bars. Miezin, McGuinness and Allman (1982) tested the response of cells to a moving bar against a random dot background. When a bar moving in the cell's preferred direction was presented against a background of random dots moving in the same direction, the response was greatly reduced, relative to the response of the cell to a bar moving against a background of stationary dots. When the background was moved in the opposite direction, the response of the unit to the moving bar was either less inhibited, or in some cases strongly facilitated. The effect of the background pattern was stronger when it extended beyond the cell's conventionally defined receptive field; MT units could be influenced by movement presented at large distances from the center of their receptive field. The effect of the surround was found to be directionally selective; a preferred direction of movement presented in the center of a cell's receptive field would result in maximum inhibition of the cell's response when presented in the surround. Miezin *et al.* (1982) argue that an antagonistic center-surround organization of directionally selective mechanisms provides a comparison between local movement in the visual field, and that of surrounding areas, which may be useful for the separation of figure from ground. This idea is very similar to that developed by Reichardt and Poggio (1979) for relative movement detection in the fly.

### 8.3.4. Computational Considerations for Area MT

In this section, we discuss the possible role of area MT in the integration of directionally selective measurements from units in striate cortex, to compute the velocity field and its discontinuities. The direct projection from striate cortex to MT places this area at the right location in the visual pathway to perform this next stage in the analysis of motion. The topographical organization of receptive fields suggests that MT could provide a retinotopic representation of movement in the changing image. In the discussion that follows, specific questions that arise from computational considerations are italicized.

#### The Velocity Field Computation

A precise and detailed velocity field, such as that described in earlier chapters, is probably computed only in the central visual field. It is appropriate that a large part of MT is devoted to a representation of this area. A precise measurement of motion is needed primarily for the computation of three-dimensional structure from motion; the ability of the human visual system to perform this computation appears to degrade quickly as distance from the center of the visual field increases (Shimon Ullman, personal communication). One might expect this to be the case, because the coarseness of the initial filters in the human visual system rules out a detailed computation of velocity in the periphery. We instead expect only a rough computation of motion in the periphery, perhaps concerned primarily with the gross translational motion that occurs with movement of the observer. The detection of motion discontinuities is important throughout the visual field, but may be particularly important in the periphery, where the fast detection of sudden movements serves as an early warning system for approaching objects.

One set of questions for neurophysiological investigation concerns the nature of the input to MT. *What are the properties of units in striate cortex that project to MT? For example, does this projection arise largely from directionally selective cortical cells?* It would at least be of interest to know the general properties of units in layers IVb and VI, which appear to provide the primary inputs to MT.

An important property for units involved in the integration of initial motion measurements is the dissociation of orientation and direction of motion. Ideally such units would exhibit a specificity for the two-dimensional direction of movement of a stimulus, independent of its spatial structure. This property was observed, to some extent, in area MT, particularly in the studies of Movshon *et al.* (1983). Of particular interest would be the nature of the directional selectivity of the output cells of area MT. Neurophysiological studies that use moving spots and bars of light as stimuli yield fairly broad direction tuning curves for MT neurons. Albright and Gross (1982) observed a wide range of directional tuning, from bandwidths of about 25° to 270°. If a precise velocity field is only computed in the central visual field, sharper direction tuning might be expected there than in the periphery. *Does a large variation in selectivity for direction of motion exist everywhere in the visual field, or is there a sharper direction tuning for units whose receptive fields are located in the central visual field?* Zeki (1974) suggests that central units in MT of the rhesus monkey may be more sharply tuned.

It should be noted that there are at least two ways in which the visual system can represent the direction of velocity. First, the system can use narrowly tuned mechanisms that respond only to a small range of velocity directions. Second, it can use broadly tuned units, and represent velocity direction implicitly through the relative responses of a set of units. Broad directional selectivities for units in MT would not rule out a role in the velocity field computation.

It is possible that a moving spot or bar is not sufficiently rich in information to elicit a specific response from MT units, either because of its small size, or lack of a range of feature orientations. The random dot patterns used by Baker *et al.* (1981), and the patterns of superimposed sine wave gratings used by Movshon *et al.* (1983), may be more effective stimuli for MT units. *Is the direction tuning curve for these richer, two-dimensional patterns the same as that for a moving spot or bar?*

A detailed velocity field computation requires the representation of both direction and magnitude of velocity. Baker *et al.* (1981) found a range of optimal speeds to which MT units respond. One might expect there to be a greater sensitivity to lower speeds in the central visual field, due to the small size of the initial filters, and a selectivity for higher velocities in the periphery. *Are optimal speeds correlated with receptive field location, or is there a wide range of optimal speeds for MT units within a single area of the visual field?* The study by Maunsell and Van Essen (1983) suggests that selectivity for a wide range of speeds may exist in all parts of the visual field, although a greater proportion of peripheral neurons appear to be sensitive to higher speeds. One can also ask about the speed selectivities of the inputs to area MT, an issue addressed briefly in Section 8.2.

**The Detection of Motion Discontinuities**

The study of Miezin, McGuinness and Allman (1982) clearly indicates the preference of some MT cells for relative movement. In Chapter 5, we discussed the detection of discontinuities via a comparison between the sign of direction of motion of nearby zero-crossing segments of the same orientation. We mentioned both the direct comparison between a small number of zero-crossing motion detectors, and a center-surround comparison of a large number of detectors. The experiments of Miezin *et al.* (1982) suggest that a center-surround organization of directionally-selective units may be used for the detection of relative movement.

It would be interesting to test the response of relative movement detectors to a discontinuity that is defined purely by differential movement, such as those created in the dynamic random dot patterns studied in psychophysical experiments. If these units respond to such a discontinuity, it can then be asked: *What is the difference in direction of velocity across the discontinuity required to elicit a significant response? Does the unit respond when the discontinuity is defined by a difference in magnitude of velocity alone? If so, how much of a difference in magnitude is required to elicit a significant response? How selective is the unit for the size of a patch that is moving relative to its surround? Is there a variation with eccentricity, of the units' dependence on any of the above properties?* Answers to these questions might provide some clue to the underlying algorithm for discontinuity detection.

An important question for discontinuity detection concerns the input representation for the underlying computations. *Do the inputs to these relative movement units represent only the component of motion in the direction perpendicular to features in the image, or do they represent real two-dimensional velocities?* In Chapter 5, we noted that if the detection of discontinuities were based on initial motion measurements along zero-crossing contours, the comparison between zero-crossing motion detectors of similar orientation is required. Miezin *et al.* (1982) found the center-surround mechanism to be directionally selective. *Does this mechanism also exhibit a selectivity for orientation?* One of the consequences of basing the discontinuity detection on initial measurements is that, if the structure of the pattern is such that there exists only a few zero-crossings at the necessary orientations to detect the discontinuity, detection should be hindered. One could test MT units with random dot patterns that have been filtered so as to reduce the presence of zero-crossings at orientations required to detect a particular discontinuity.

It is also possible to detect motion discontinuities through the direct comparison of the sign of direction of motion of nearby zero-crossings of similar orientations. If such a scheme were implemented in MT, one might expect to find units that respond to a bar whose edges are moving in opposite directions.

# 9. SUMMARY AND CONCLUSIONS

In this book, we have investigated two problems in the early analysis of visual motion. The main focus has been the computation of the two-dimensional velocity field that results from the changing projection of three-dimensional surfaces in motion. Related to this computation is the detection of discontinuities in the velocity field. A computational approach to these problems provided us with a framework within which the tools of mathematics, physics, neurobiology and perceptual psychology could be combined. Each of these disciplines provides important and different insights into the computations underlying the measurement and use of visual motion; in isolation, however, no single discipline can provide a full understanding of these computations at all of the important levels: the computational theory, algorithm and implementation.

A theory of the velocity field computation has been developed, with three main components. First, initial measurements of motion in the image take place at the location of significant intensity changes, which give rise to zero-crossings in the output of the convolution of the image with a $\nabla^2 G$ operator. The initial motion measurements provide the component of velocity in the direction perpendicular to the local orientation of the zero-crossing contours. Second, these initial measurements are integrated along contours to compute the two-dimensional velocity field. Third, an additional constraint of smoothness of the velocity field, based on the physical constraint that surfaces are generally smooth, allows the computation of a unique velocity field. Regarding the detection of discontinuities, the main conclusion is that this detection can be based on the initial motion measurements, and therefore may begin to take place prior to the integration of these measurements for the full two-dimensional velocity field computation.

There are three aspects of this work that are of fundamental importance. First, it has been shown that an additional constraint for the velocity field computation can be formulated that allows the analysis of general motion, while providing a velocity field solution that is unique, and physically plausible. Second, the velocity field and discontinuity computations developed here lead naturally to algorithms that are biologically feasible, in that they require only simple, local, parallel operations.

Finally, the proposed velocity field computation appears to yield solutions that are consistent with human motion perception. Thus, the computational theory that has been put forth in this book is viable from a mathematical and physical point of view, and contributes toward our understanding of the measurement of motion in the human visual system.

# Appendix I

In this appendix, we show that if a smooth, two-dimensional curve undergoes rigid rotation and translation in the image plane, there are at least two points on the curve for which the direction of velocity is known. These two points are such that $v^{\perp}(s) = 0$, and the local velocity vector is parallel to the local tangent to the curve (their directions may differ by 0° or 180°).

**Proposition.** *Given a smooth, closed two-dimensional curve $C$, which is undergoing rigid translation and rotation in the image plane, there exists at least two points on $C$ for which $\mathbf{V}(s) = a\mathbf{T}(s)$, where $\mathbf{V}(s)$ and $\mathbf{T}(s)$ are the local velocity and tangent vectors on the curve, and $a$ is a scalar.*

**Proof.** First consider the case in which $C$ undergoes pure translation. $\mathbf{V}(s)$ is then constant. Since $C$ is closed, there must be at least two points whose tangent is parallel to $\mathbf{V}(s)$. The motion of $C$ can otherwise be described as pure rotation about a point $p$ in the image plane. In this case, $\mathbf{V}(s)$ at a point $p_i$ on the curve is directed perpendicular to $\mathbf{R}(s)$, the position vector connecting $p$ and $p_i$. To show that there exists two points at which the velocity and tangent vectors have the same orientation, it is sufficient to show that there are two points at which $\mathbf{R}(s)$ and $\mathbf{T}(s)$ are perpendicular, i.e., where $\mathbf{R}(s) \cdot \mathbf{T}(s) = 0$. Since $C$ is closed, then either $\mathbf{R}(s) \cdot \mathbf{T}(s) = 0$ everywhere, or $\mathbf{R}(s) \cdot \mathbf{T}(s)$ must take on positive and negative values, and must return to the same value, as we trace around the full curve. If the curve is smooth and continuous, then $\mathbf{R}(s) \cdot \mathbf{T}(s)$ must also be smooth and continuous. Thus, from the Intermediate Value Theorem, there must be at least two points on the curve for which $\mathbf{R}(s) \cdot \mathbf{T}(s) = 0$, that is, for which the velocity vector and tangent vector are parallel.

From the changing image, points at which the direction of velocity is parallel to the direction of the tangent may be found by locating the points at which $v^{\perp}(s) = 0$.

# Appendix II

In the first part of this appendix, we establish the uniqueness of the velocity field that satisfies the velocity constraints obtained from the changing image, and minimizes total variation, given by the measure: $\int |\frac{\partial \mathbf{V}}{\partial s}|^2 ds$. In the second part of the appendix, some simple examples are presented, that illustrate the results of other measures of variation that do not, in general, yield unique velocity field solutions.

## A. Uniqueness of the Velocity Field

In this section of the appendix, the following theorem is used to establish the uniqueness of the velocity field solution (a proof may be found in Rudin (1973)):

**Theorem**. *Suppose there exists a complete semi-norm $\Theta$ on a space of functions $H$, and that $\Theta$ satisfies the parallelogram law. Then, every nonempty closed convex set $E \subset H$ contains a unique element v of minimal norm, up to an element of the null space. Thus, the family of minimal functions is*

$$\{v + s \mid s \in S\}$$

*where*

$$S = \{w \mid v + w \in E\} \cap \mathcal{N}$$

*and $\mathcal{N}$ is the null space of the functional*

$$\mathcal{N} = \{u \mid \Theta(u) = 0\}.$$

**Proof.** Let $V$ be the space of all real two-dimensional vector-valued functions along a curve that are continuous, and have continuous first partial derivatives. We first show that the functional $\Theta(\mathbf{V}) = \{\int |\frac{\partial \mathbf{V}}{\partial s}|^2 ds\}^{\frac{1}{2}}$ is a complete semi-norm on this space that satisfies the parallelogram law. We then show that the subset $W \subset V$ of vector functions that satisfy the image constraints is convex. It then follows from the above theorem that there exists a unique element $w \in W$ of minimal norm, up to possibly an element of the null space.

To show that the functional $\Theta(\mathbf{V}) = \{\int |\frac{\partial \mathbf{V}}{\partial s}|^2 ds\}^{\frac{1}{2}}$ is a semi-norm that satisfies the parallelogram law, we take advantage of the following: If $\mu$ is a semi-inner product, defined on the space $V$, and $v \in V$, then $\Theta(v) = \mu(v, v)^{\frac{1}{2}}$ is a semi-norm on the space. A semi-inner product is a function $\mu: V \times V \mapsto \Re$ written $\mu(v, w)$ satisfying the following ($w, x, y, z \in V$):

1. $\mu(y, x) = \mu(x, y)$.
2. $\mu(x + y, z) = \mu(x, z) + \mu(y, z)$.
3. $\mu(ax, y) = a\mu(x, y)$, $a \in \Re$.
4. $\mu(x, x) \geq 0$.

The functional $\mu(v, w) = \int \left(\frac{\partial v}{\partial s} \cdot \frac{\partial w}{\partial s}\right)ds$ is a semi-inner product:

1. $\mu(y, x) = \int \left(\frac{\partial y}{\partial s} \cdot \frac{\partial x}{\partial s}\right)ds = \int \left(\frac{\partial x}{\partial s} \cdot \frac{\partial y}{\partial s}\right)ds = \mu(x, y)$.
2. $\mu(x+y, z) = \int \left(\frac{\partial(x+y)}{\partial s} \cdot \frac{\partial z}{\partial s}\right)ds = \int \left(\frac{\partial x}{\partial s} + \frac{\partial y}{\partial s}\right) \cdot \frac{\partial z}{\partial s} ds = \int \{\left(\frac{\partial x}{\partial s} \cdot \frac{\partial z}{\partial s}\right) + \left(\frac{\partial y}{\partial s} \cdot \frac{\partial z}{\partial s}\right)\}ds$

$$= \mu(x, z) + \mu(y, z).$$

3. $\mu(ax, y) = \int \left(\frac{\partial(ax)}{\partial s} \cdot \frac{\partial y}{\partial s}\right)ds = a\int \left(\frac{\partial x}{\partial s} \cdot \frac{\partial y}{\partial s}\right)ds = a\mu(x, y)$.
4. $\mu(x, x) = \int \left(\frac{\partial x}{\partial s} \cdot \frac{\partial x}{\partial s}\right)ds \geq 0$.

If $\mu(v, w)$ is a semi-inner product, then $\Theta(v) = \mu(v, v)^{\frac{1}{2}} = \{\int \left(\frac{\partial v}{\partial s} \cdot \frac{\partial v}{\partial s}\right)ds\}^{\frac{1}{2}} = \{\int \left|\frac{\partial v}{\partial s}\right|^2 ds\}^{\frac{1}{2}}$ is a semi-norm. $\Theta(v)$ also satisfies the parallelogram law:

$[\Theta(v + w)]^2 + [\Theta(v - w)]^2 =$

$$= \int \left(\frac{\partial(v + w)}{\partial s} \cdot \frac{\partial(v + w)}{\partial s}\right)ds + \int \left(\frac{\partial(v - w)}{\partial s} \cdot \frac{\partial(v - w)}{\partial s}\right)ds$$

$$= 2\int \left(\frac{\partial v}{\partial s} \cdot \frac{\partial v}{\partial s}\right)ds + 2\int \left(\frac{\partial w}{\partial s} \cdot \frac{\partial w}{\partial s}\right)ds$$

$$= 2[\Theta(v)]^2 + 2[\Theta(w)]^2.$$

It now remains to show that the semi-norm is complete. To do this, it must be shown that every Cauchy sequence $\{\mathbf{V}_n\}$, where $\mathbf{V}_n \in V$, converges under this semi-norm. We begin with the functional $\Theta(\mathbf{V}) = \{\int |\frac{\partial \mathbf{V}}{\partial s}|^2 ds\}^{\frac{1}{2}}$, where $\Theta: V \mapsto \Re$. Let $\mathbf{f}(\mathbf{V}) = \frac{\partial \mathbf{V}}{\partial s}$, and define a new functional $\Phi(\mathbf{f}) = \{\int |\mathbf{f}|^2 ds\}^{\frac{1}{2}}$, with $\Phi: \partial V \mapsto \Re$, where $\partial V$ denotes the space of all partial derivatives of elements of $V$. $\Phi(\mathbf{f})$ is an $L^2$ norm, and is therefore complete on $\partial V$ (for a proof, see Rudin 1966, p. 66). Let $\{\mathbf{V}_n\}$ be a Cauchy sequence of two-dimensional vector functions in $V$, under the norm $\Theta$; then the sequence $\{\frac{\partial \mathbf{V}_n}{\partial s}\}$ is a Cauchy sequence in $\partial V$, under the norm $\Phi$.

Since $\Phi$ is complete in $\partial V$, then any Cauchy sequence in $\partial V$ must converge under $\Phi$; that is, there exists an element $\mathbf{f} \in \partial V$ such that:

$$\lim_{n\to\infty} \Phi\left(\frac{\partial \mathbf{V}_n}{\partial s} - \mathbf{f}\right) = 0.$$

Now, let $\mathbf{V} = \int \mathbf{f} ds$. $V$ contains only those vector functions along a curve that are continuous, with continuous first partial derivatives; it follows that $\mathbf{f}$ and $\mathbf{V}$ are continuous, so that $\mathbf{V} \in V$. ($\mathbf{V}$ may not be unique, but this is to be expected, because the functional $\Theta$ is only a semi-norm.) $\mathbf{V}$ satisfies the following:

$$\lim_{n\to\infty} \Theta(\mathbf{V}_n - \mathbf{V}) = 0.$$

Therefore the Cauchy sequence $\{\mathbf{V}_n\}$ converges, and the semi-norm is complete.

The second statement that must be shown is that the subset of velocity fields satisfying the image constraints is convex. Let $W \subset V$ be the subset of $V$ that satisfies the constraints derived from the image. These constraints are as follows:

1. $v^{\perp}(s)$ is known everywhere along the curve.

2. There may exist a set of points $\{p_i\}$ on $C$ for which $v^{\perp}(s)$ and $v^{\top}(s)$ are known.

3. There may exist a set of points $\{p_i\}$ for which $\frac{v^{\perp}(s)}{v^{\top}(s)}$ is known.

If $W$ is convex, it must be the case that:

$$tw_1 + (1-t)w_2 \in W, \text{ for all } w_1, w_2 \in W, \text{ and } 0 \le t \le 1.$$

For convenience of notation, we write $v^{\perp}$ and $v^{\top}$ for $v^{\perp}(s)$ and $v^{\top}(s)$, and let $v\mid_{p_i}$ denote the value of the function $v$ at the point $p_i$.

1. If $w_1, w_2 \in W$, then $w_1^{\perp}\mid_{p_i} = w_2^{\perp}\mid_{p_i} = v^{\perp}\mid_{p_i}$, hence:

$$[tw_1 + (1-t)w_2]^{\perp}\mid_{p_i} = tw_1^{\perp}\mid_{p_i} + (1-t)w_2^{\perp}\mid_{p_i} = v^{\perp}\mid_{p_i}.$$

2. Similarly, if $w_1, w_2 \in W$, then $w_1^{\top}\mid_{p_i} = w_2^{\top}\mid_{p_i} = v^{\top}\mid_{p_i}$, hence:

$$[tw_1 + (1-t)w_2]^{\top}\mid_{p_i} = tw_1^{\top}\mid_{p_i} + (1-t)w_2^{\top}\mid_{p_i} = v^{\top}\mid_{p_i}.$$

3. If the direction of velocity at the point $p_i$ is the same for $w_1$ and $w_2$, then any linear combination of $w_1$ and $w_2$ yields a velocity vector at $p_i$ that is parallel to this direction.

Finally, if $w_1$ and $w_2$ are continuous, with continuous first partial derivatives, then this will also be true of $[tw_1 + (1 - t)w_2]$. Therefore, the subset $W \subset V$, which satisfies the image constraints, is convex. This completes the proof.

## B. Other Measures of Variation

In this section, we first present an example for which the measures of variation given by $\int |\frac{\partial \varphi}{\partial s}| ds$ (variation in direction) and $\int \frac{\partial |\mathbf{V}|}{\partial s} ds$ (variation in magnitude) do not yield unique solutions. We then present an example for which a velocity field that minimizes $\int |\frac{\partial \mathbf{V}}{\partial s}| ds$ does not exist. In both cases, the formulation of velocity constraints in velocity space is used to analyze the behavior of the smoothness measures.

### B.1 A Line Segment in Motion

Consider a simple line segment that moves rigidly in the image plane, as shown in Figure A.1a. Let us assume that the perpendicular component of velocity along the line, $v^{\perp}(s)$, and the velocities of the two endpoints, $\mathbf{V}(s_o)$ and $\mathbf{V}(s_n)$, are known. The velocity of a particular point, $\mathbf{V}(s_i)$ on the line is constrained to project to a line $\mathbf{L}_i$ in velocity space, whose orientation is equal to the orientation of the line in the image, and whose perpendicular distance from the origin in velocity space is given by $v^{\perp}(s_i)$. This constraint is illustrated in Figure A.1b. A set of evenly spaced points along the line yields a set of constraint lines in velocity space, which are evenly spaced from the origin, as shown in Figure A.1c. In addition, the velocity vectors associated with the endpoints are shown in velocity space.

The true velocity field for the moving line segment corresponds to the straight line $l$ shown in Figure A.2a, connecting the two known velocity vectors. The velocity of a point on the line segment, $\mathbf{V}(s_i)$, is given by the intersection of $l$ and $\mathbf{L}_i$. Any valid velocity field must correspond to a path in velocity space that begins at the endpoint of $\mathbf{V}(s_o)$, intersects each constraint line in the order that they arise along the image curve, and ends at the endpoint of $\mathbf{V}(s_n)$. Three examples of valid velocity fields correspond to the curves shown in Figure A.2b. Computation of the velocity field in this case involves finding the valid velocity field that exhibits the least variation, given one of the measures of variation.

Suppose we want to find the velocity field that is consistent with the constraints shown in Figure A.1c, and minimizes the total change in direction of velocity, given

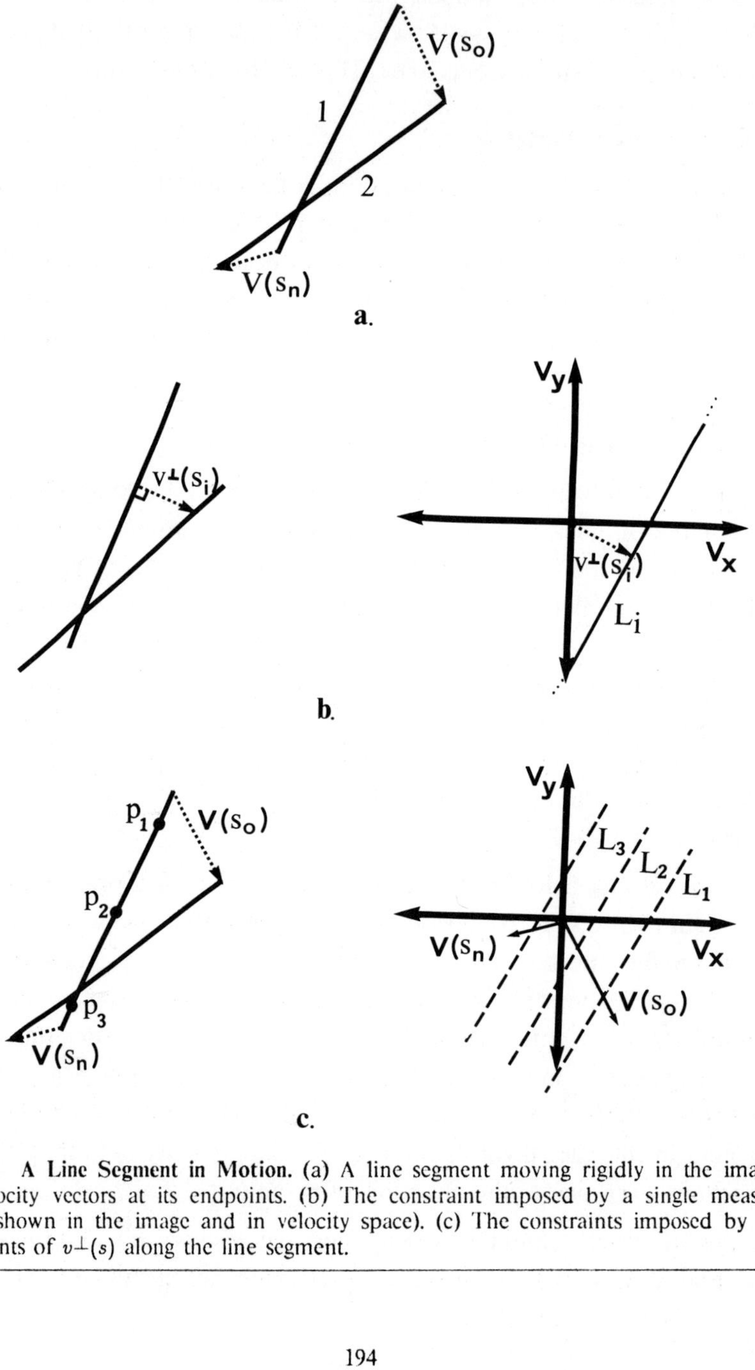


**Figure A.1. A Line Segment in Motion.** (a) A line segment moving rigidly in the image, with known velocity vectors at its endpoints. (b) The constraint imposed by a single measurement of $v^{\perp}(s)$ (shown in the image and in velocity space). (c) The constraints imposed by multiple measurements of $v^{\perp}(s)$ along the line segment.

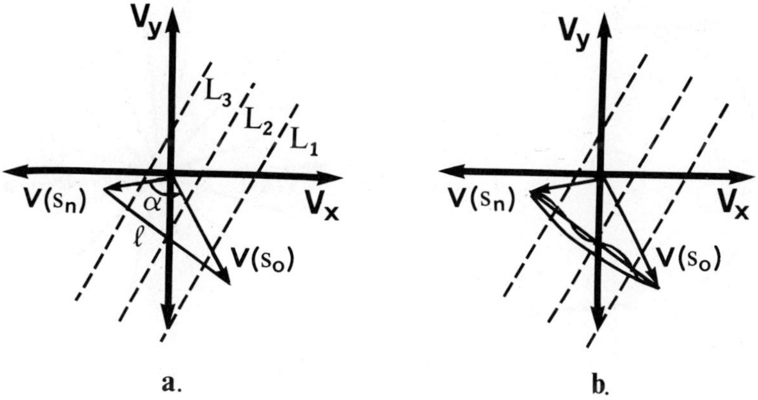


**Figure A.2. Velocity Fields for a Line Segment in Motion.** (a) The true velocity field in the image projects to the line $l$ in velocity space. (b) The curved paths correspond to other valid velocity field solutions.

by: $\int |\frac{\partial \varphi}{\partial s}| ds$. For the case of the true velocity field, $\varphi$ increases monotonically; its total change is given by the angle $\alpha$, shown in Figure A.2a; hence $\int |\frac{\partial \varphi}{\partial s}| ds = \alpha$. $\alpha$ is the smallest total variation in direction that a velocitv field consistent with these constraints may obtain. There are other velocity fields, however, that distribute the total change $\alpha$ over the curve monotonically and smoothly, in a way that is consistent with $v^{\perp}(s)$. Other examples correspond to the curved paths in Figure A.2b; for these velocity fields, it is also the case that $\int |\frac{\partial \varphi}{\partial s}| ds = \alpha$. Hence, in the simple case of a line segment moving rigidly in the image plane, minimizing the total change in direction of velocity alone does not yield a unique velocity field solution.

Now consider minimizing the variation in magnitude of velocity, given by the measure: $\int \frac{\partial |\mathbf{V}|}{\partial s} ds$. Let $m_1$ and $m_2$ denote the magnitudes of the velocity vectors associated with the endpoints. It then follows that $\int \frac{\partial |\mathbf{V}|}{\partial s} ds \geq |m_1 - m_2|$. Any velocity field that is consistent with the image constraints, and changes monotonically in magnitude along the line will yield $\int \frac{\partial |\mathbf{V}|}{\partial s} ds = |m_1 - m_2|$. This solution is again not unique.

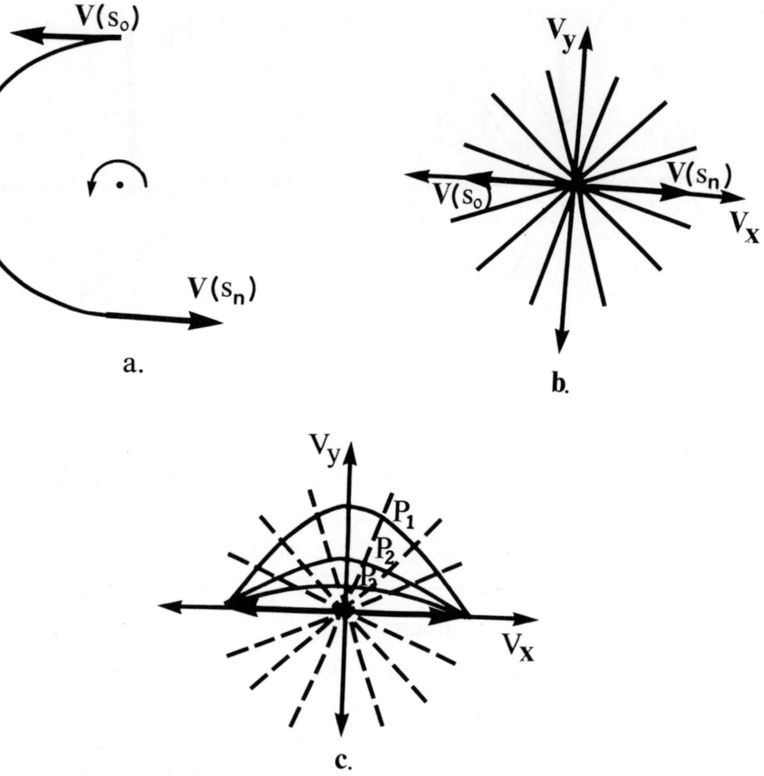


**Figure A.3. A Rotating Semi-circle.** (a) A semi-circle rotating in the image about its center, with known velocity vectors at its endpoints. (b) The constraint imposed by the measurements of $v^{\perp}(s)$, shown in velocity space. (c) The solid curves correspond to valid velocity fields.

## B.2 A Rotating Semi-circle

As a second example, consider a semi-circle rotating about its center, as shown in Figure A.3a. Assume that the velocity vectors $\mathbf{V}(s_\circ)$ and $\mathbf{V}(s_n)$ at the endpoints are known. Along the contour, $v^{\perp}(s) = 0$, so these measurements give rise to lines of constraint in velocity space that intersect at the origin, as shown in Figure A.3b. The known velocity vectors at the endpoints are also shown in Figure A.3b. Consider the measure of variation given by: $\int |\frac{\partial \mathbf{V}}{\partial s}| ds$. As before, a valid velocity field must correspond to a path in velocity space that begins at the endpoint of $\mathbf{V}(s_\circ)$, intersects the lines of constraint in the order that they arise from the image curve, and ends at the endpoint of $\mathbf{V}(s_n)$. The value of the measure $\int|\frac{\partial \mathbf{V}}{\partial s}|ds$ is equal to the length of the path in velocity space. Thus, a velocity field that minimizes this measure must give rise to a path of minimum length in velocity space. For this particular example, however, a minimum length path satisfying the constraints does not exist. In Figure A.3c, we show three paths in velocity space, labeled $P_1$, $P_2$ and $P_3$, that correspond to valid velocity fields. As we move from $P_1$ to $P_3$, the length of the path decreases. Thus, the value of the measure $\int|\frac{\partial \mathbf{V}}{\partial s}|ds$ decreases, for the corresponding velocity fields. We can continue indefinitely, however, to find paths of shorter length that satisfy the constraints. The paths will approach the horizontal line, but can never reach it exactly, because the horizontal line does not correspond to a valid velocity field. Thus, a velocity field of least variation, given this particular measure of variation, does not exist.

# Appendix III

In this appendix, we show the following:

**Proposition**. *Suppose that a rigid three-dimensional object, consisting of straight lines intersecting in space, projects onto the image plane, using orthographic projection, in such a way that line intersections are preserved (that is, two lines intersect in the image if and only if their generators intersect in space). Further, suppose that this object rotates in space with angular velocity* $\omega$ *about a single axis* $\mathbf{n} = [n_x, n_y, n_z]^T$. *Then the two-dimensional velocity field that satisfies* $v^{\perp}(s)$ *measured along lines in the image and minimizes* $\int |\frac{\partial \mathbf{V}}{\partial s}|^2 ds$, *is the correct two-dimensional projected velocity field.*

We divide the proof into two parts: the first shows that the constraint imposed by two lines meeting at a vertex is sufficient to compute the correct two-dimensional velocity of the vertex; the second shows that if a rigid line segment is moving in space, and the projected two-dimensional velocity vectors at its endpoints are known, then the velocity field satisfying $v^{\perp}(s)$ along the projection of the line onto the image, which minimizes $\int |\frac{\partial \mathbf{V}}{\partial s}|^2 ds$, is the correct projected velocity field. These results are extensions of some previous results derived by Alan Yuille (personal communication). Throughout this appendix, the argument, $s$, to functions that vary along a curve, has been omitted.

**Proposition (a)**. *Suppose that two lines* $l_1$ *and* $l_2$ *in space, intersect at the point* $(x_1, y_1, z_1)$, *and the configuration rotates rigidly with angular velocity* $\omega$ *about a single axis in space, given by* $\mathbf{n} = [n_x, n_y, n_z]^T$. *Assume that* $l_1$ *and* $l_2$ *project onto lines* $l_1^p$ *and* $l_2^p$ *in the image plane, using orthographic projection, which have different orientations. Then the constraint provided by* $v^{\perp}$ *on* $l_1^p$ *and* $l_2^p$ *is sufficient to compute the correct projected two-dimensional velocity at* $(x_1, y_1, z_1)$.

**Proof**. First, let $\alpha$ and $\beta$ denote the orientations of $l_1^p$ and $l_2^p$ in the image plane. The true velocity field, $\mathbf{V}$, is given by:

$$\mathbf{V} = \omega z \begin{bmatrix} n_y \\ -n_x \end{bmatrix} + \omega n_z \begin{bmatrix} -y \\ x \end{bmatrix}.$$

Thus, $\mathbf{V} = (\omega z n_y - \omega y n_z, -\omega z n_x + \omega x n_z)$. Let $\mathbf{u}^{\top}$ and $\mathbf{u}^{\perp}$ denote the unit tangent and normal vectors, and $v^{\top}$ and $v^{\perp}$ denote the tangential and normal components of velocity. Then:

$$\mathbf{V} = v^{\top}\mathbf{u}^{\top} + v^{\perp}\mathbf{u}^{\perp}.$$

For $l_1^p$, we have the following:

$$\begin{aligned}
\mathbf{u}^{\top} &= (\cos\alpha,\ \sin\alpha)\\
\mathbf{u}^{\perp} &= (\sin\alpha,\ -\cos\alpha)\\
v^{\top} &= \mathbf{V}\cdot\mathbf{u}^{\top} = (\omega z n_y - \omega y n_z)\cos\alpha + (\omega x n_z - \omega z n_x)\sin\alpha\\
v^{\perp} &= \mathbf{V}\cdot\mathbf{u}^{\perp} = (\omega z n_y - \omega y n_z)\sin\alpha - (\omega x n_z - \omega z n_x)\cos\alpha.
\end{aligned}$$

For $l_2^p$, we have the following:

$$\begin{aligned}
\mathbf{u}^{\top} &= (\cos\beta,\ \sin\beta)\\
\mathbf{u}^{\perp} &= (\sin\beta,\ -\cos\beta)\\
v^{\top} &= \mathbf{V}\cdot\mathbf{u}^{\top} = (\omega z n_y - \omega y n_z)\cos\beta + (\omega x n_z - \omega z n_x)\sin\beta\\
v^{\perp} &= \mathbf{V}\cdot\mathbf{u}^{\perp} = (\omega z n_y - \omega y n_z)\sin\beta - (\omega x n_z - \omega z n_x)\cos\beta.
\end{aligned}$$

It is assumed that $v^{\perp}$ is known, and that the tangential component $v^{\top}$ is unknown. The velocity of the point $(x_1, y_1, z_1)$ must satisfy constraints imposed by both lines $l_1^p$ and $l_2^p$. Let $c_1$ and $c_2$ denote the unknown $v^{\top}$ along the lines $l_1^p$ and $l_2^p$ at the point $(x_1, y_1, z_1)$. There are then two equations for the velocity at this point, in terms of the two unknowns $c_1$ and $c_2$. From $l_1^p$ we have:

$$\mathbf{V} = [(\omega z_1 n_y - \omega y_1 n_z)\sin\alpha - (\omega x_1 n_z - \omega z_1 n_x)\cos\alpha]\begin{bmatrix}\sin\alpha\\ -\cos\alpha\end{bmatrix} + c_1\begin{bmatrix}\cos\alpha\\ \sin\alpha\end{bmatrix}.$$

From $l_2^p$:

$$\mathbf{V} = [(\omega z_1 n_y - \omega y_1 n_z)\sin\beta - (\omega x_1 n_z - \omega z_1 n_x)\cos\beta]\begin{bmatrix}\sin\beta\\ -\cos\beta\end{bmatrix} + c_2\begin{bmatrix}\cos\beta\\ \sin\beta\end{bmatrix}.$$

These two equations are satisfied for:

$$c_1 = (\omega z_1 n_y - \omega y_1 n_z)\cos\alpha + (\omega x_1 n_z - \omega z_1 n_x)\sin\alpha$$

$$c_2 = (\omega z_1 n_y - \omega y_1 n_z)\cos\beta + (\omega x_1 n_z - \omega z_1 n_x)\sin\beta.$$

$c_1$ and $c_2$ are equivalent to the true $v^{\top}$ stated previously; therefore, the correct projected two-dimensional velocity has been obtained at the point $(x_1, y_1, z_1)$.

**Proposition (b).** *Suppose a line segment $l$ in space rotates rigidly with angular velocity $\omega$ about a single axis, given by $\mathbf{n} = [n_x, n_y, n_z]^T$. Let $l^p$ denote the orthographic projection of $l$ onto the image plane. Further, suppose that the projected two-dimensional velocity vectors at the endpoints, denoted by $\mathbf{V}_0$ and $\mathbf{V}_1$, are known. Then the two-dimensional velocity field that satisfies the constraint given by $\mathbf{V}_0$, $\mathbf{V}_1$ and $v^{\perp}$ along $l^p$, and minimizes $\int |\frac{\partial \mathbf{V}}{\partial s}|^2 ds$, is the correct projected two-dimensional velocity field.*

**Proof.** First, $\mathbf{u}^{\perp}$ and $\mathbf{u}^{\top}$ are constant along the line segment; therefore:

$$\frac{\partial \mathbf{V}}{\partial s} = \frac{\partial v^{\perp}}{\partial s}\mathbf{u}^{\perp} + \frac{\partial v^{\top}}{\partial s}\mathbf{u}^{\top}.$$

Total variation of velocity along the segment is then given by:

$$\int_{s_0}^{s_1} \left|\frac{\partial \mathbf{V}}{\partial s}\right|^2 ds = \int_{s_0}^{s_1} \left[\left(\frac{\partial v^{\perp}}{\partial s}\right)^2 + \left(\frac{\partial v^{\top}}{\partial s}\right)^2\right] ds$$

where $s_0$ and $s_1$ denote arclength at the two endpoints. Since $\frac{\partial v^{\perp}}{\partial s}$ is known, the above expression is minimized when the second term is minimized:

$$\int_{s_0}^{s_1} \left(\frac{\partial v^{\top}}{\partial s}\right)^2 ds.$$

Let $v_0^{\top}$ and $v_1^{\top}$ denote the known tangential components of velocity at the endpoints of $l$. We want to minimize the above functional, subject to these known velocities. The problem can be formulated as an *Isoperimetric* problem, and the Euler-Lagrange equations can be used to obtain a solution. The general problem can be stated as follows:

$$\text{minimize} \quad I = \int_{s_0}^{s_1} \mathbf{F}\left[v^{\top}, \frac{\partial v^{\top}}{\partial s}, s\right] ds$$

$$\text{subject to} \quad \int_{s_0}^{s_1} \Phi\left[v^{\top}, \frac{\partial v^{\top}}{\partial s}, s\right] ds = c_k.$$

In this case, we seek to minimize:

$$\int_{s_0}^{s_1} \left(\frac{\partial v^{\top}}{\partial s}\right)^2 ds$$

subject to the following constraint:

$$\int_{s_0}^{s_1} \left(\frac{\partial v^{\top}}{\partial s}\right) ds = v_1^{\top} - v_0^{\top}.$$

The general solution is given by the Euler-Lagrange equation:

$$\frac{d}{ds}\left(\frac{\partial \Psi}{\partial \left(\frac{\partial v^{\top}}{\partial s}\right)}\right) - \frac{\partial \Psi}{\partial s} = 0$$

where $\Psi = \mathbf{F} + \mu\Phi$. $\mu$ denotes the Lagrange multiplier. In this case:

$$\mathbf{F} = \left(\frac{\partial v^{\top}}{\partial s}\right)^2, \qquad \Phi = \frac{\partial v^{\top}}{\partial s}.$$

We therefore have the following differential equation:

$$\frac{d}{ds}\left[2\left(\frac{\partial v^{\top}}{\partial s}\right) + \mu\right] = 0$$

which leads to the equation:

$$\frac{\partial^2 v^{\top}}{\partial s^2} = 0.$$

The solution is therefore: $v^{\top} = as + b$. From the known tangential components of velocity at the endpoints, the following solution is obtained:

$$v^{\top} = \frac{v_1^{\top} - v_0^{\top}}{s_1 - s_0}(s - s_0) + v_0^{\top}.$$

Since the true $v^{\top}$ and computed tangential components are both linear functions that satisfy the same endpoint velocities, they must be the same function. Therefore the smoothest velocity field is the correct projected two-dimensional velocity field.

The original statement of the problem now follows straightforwardly from these two results. In addition, we have the following corollary:

> **Corollary**. *Suppose a simple polygon is rotating rigidly in the image plane. Then the two-dimensional velocity field that satisfies* $v^{\perp}$ *measured along line segments in the image, and minimizes* $\int \left|\frac{\partial \mathbf{V}}{\partial s}\right|^2 ds$, *is the correct two-dimensional velocity field.*

Finally, the above results also hold for an additional translation of the object in space.

# Appendix IV

In Chapter 4, we indicated the way in which three standard algorithms, Jacobi and Gauss-Seidel relaxation, and the method of conjugate gradients, could be used to compute the velocity field. In this appendix, we show the details of the application of these algorithms to this computation. In the first section of the appendix, the Jacobi and Gauss-Seidel relaxation methods are briefly outlined, and applied to the case in which constraints imposed by the measurements of $v^{\perp}(s)$ are to be satisfied exactly. Two different decompositions of $\mathbf{V}(s)$ are considered: first, the algorithm is formulated in such a way that the tangential components of velocity, $v^{\top}(s)$, are computed explicitly; second, a formulation is presented in which the $x$ and $y$ components are computed explicitly. In the second section of the appendix, the method of conjugate gradients is described, and applied to the velocity field computation, for the case in which the constraints imposed by the measurements of $v^{\perp}(s)$ are to be satisfied approximately. Further discussion of iterative mehods such as these can be found in Luenberger (1973), Dahlquist and Bjorck (1974), Gladwell and Wait (1979), and Wright (1982).

## A. Jacobi and Gauss-Seidel Relaxation

In Section 4.2, we suggested the use of Gauss-Seidel or Jacobi relaxation, for the case in which the constraints imposed by the measurements of $v^{\perp}(s)$ are to be satisfied exactly. This computation required the solution of a system of linear equations. Suppose the set of linear equations is given by:

$$\sum_{j=1}^{n} a_{ij} v_j = f_i, \qquad 1 \leq i \leq n$$

where the coefficients $a_{ij}$ and values $f_i$ are known. First, the equations can be rewritten as:

$$a_{ii} v_i = - \sum_{\substack{1 \leq j \leq n \\ j \neq i}} a_{ij} v_j + f_i, \qquad 1 \leq i \leq n.$$

Iterative methods compute a sequence of approximations $\mathbf{V}^{(1)}, \mathbf{V}^{(2)}$, ... to the exact solution $\mathbf{V}$, given an initial approximation $\mathbf{V}^{(0)}$, which is usually $\mathbf{V}^{(0)} = 0$.

For Jacobi relaxation, the iterative equation is given by:

$$a_{ii}v_i^{(k+1)} = -\sum_{\substack{1\leq j\leq n\\ j\neq i}} a_{ij}v_j^{(k)} + f_i, \qquad 1\leq i\leq n.$$

In the case of Gauss-Seidel relaxation, this equation becomes:

$$a_{ii}v_i^{(k+1)} = -\sum_{j=1}^{i-1} a_{ij}v_j^{(k+1)} - \sum_{j=i+1}^{n} a_{ij}v_j^{(k)} + f_i, \qquad 1\leq i\leq n.$$

The Jacobi method is inherently parallel, because the new approximation $\mathbf{V}^{(k+1)}$ is only a function of the previous approximation $\mathbf{V}^{(k)}$, while in the Gauss-Seidel method, the new approximation depends on previously computed components of the new approximation.

## A.1 The Explicit Computation of $v^\top(s)$

To compute the tangential components of velocity explicitly, we begin with the following decomposition:

$$\mathbf{V}(s) = v^\top(s)\mathbf{u}^\top(s) + v^\perp(s)\mathbf{u}^\perp(s).$$

$\frac{\partial \mathbf{V}}{\partial s}$ is given by:

$$\frac{\partial \mathbf{V}}{\partial s} = \frac{\partial v^\top}{\partial s}\mathbf{u}^\top + \frac{\partial v^\perp}{\partial s}\mathbf{u}^\perp + v^\top\frac{\partial \mathbf{u}^\top}{\partial s} + v^\perp\frac{\partial \mathbf{u}^\perp}{\partial s}.$$

From this, an expression for $\left|\frac{\partial \mathbf{V}}{\partial s}\right|^2$ can be derived:

$$\left|\frac{\partial \mathbf{V}}{\partial s}\right|^2 = \left(\frac{\partial \mathbf{V}}{\partial s}\cdot\frac{\partial \mathbf{V}}{\partial s}\right)$$

$$= \left(\frac{\partial v^\top}{\partial s}\right)^2 + \left(\frac{\partial v^\perp}{\partial s}\right)^2 + 2\{v^\perp\frac{\partial v^\top}{\partial s} - v^\top\frac{\partial v^\perp}{\partial s}\}\left(\mathbf{u}^\top\cdot\frac{\partial \mathbf{u}^\perp}{\partial s}\right)$$

$$+(v^\top)^2\left(\frac{\partial \mathbf{u}^\top}{\partial s}\cdot\frac{\partial \mathbf{u}^\top}{\partial s}\right) + (v^\perp)^2\left(\frac{\partial \mathbf{u}^\perp}{\partial s}\cdot\frac{\partial \mathbf{u}^\perp}{\partial s}\right).$$

$\left|\frac{\partial \mathbf{V}}{\partial s}\right|^2$ is an expression of the form:

$$\left(\frac{\partial v^\top}{\partial s}\right)^2 + A\frac{\partial v^\top}{\partial s} + B(v^\top)^2 + C(v^\top) + D$$

where A, B, C and D are functions of the known information, $v^{\perp}(s)$, $\mathbf{u}^{\perp}(s)$ and $\mathbf{u}^{\top}(s)$. The goal of the velocity field computation is to determine the tangential components $v^{\top}(s)$ that minimize:

$$\int \left[ \left( \frac{\partial v^{\top}}{\partial s} \right)^2 + A \frac{\partial v^{\top}}{\partial s} + B(v^{\top})^2 + C(v^{\top}) \right] ds.$$

A discrete approximation to this integral can be obtained, using finite differences. Let $v_i^{\top}$ denote the tangential component of velocity, at a point $(x_i, y_i)$ on the curve. We utilize the discrete difference $(v_i^{\top} - v_{i-1}^{\top})$ to express the partial derivative of $v^{\top}(s)$. Let $a_i$, $b_i$ and $c_i$ denote the values of the coefficients $A$, $B$ and $C$ at the point $(x_i, y_i)$. Assume for now that the curve is closed, and discrete samples are evenly spaced along the curve. A discrete approximation to the continuous functional can then be given by:

$$\Phi_4 = \sum_{i=2}^{n} \left[ (v_i^{\top} - v_{i-1}^{\top})^2 + a_i(v_i^{\top} - v_{i-1}^{\top}) + b_i(v_i^{\top})^2 + c_i v_i^{\top} \right]$$

$$+ (v_1^{\top} - v_n^{\top})^2 + a_i(v_1^{\top} - v_n^{\top}) + b_1(v_1^{\top})^2 + c_1 v_1^{\top}.$$

The function $v^{\top}(s)$ that minimizes the discrete function $\Phi_4$ satisfies the system of linear equations given by:

$$\frac{\partial \Phi_4}{\partial v_i^{\top}} = 0, \qquad 1 \leq i \leq n.$$

These equations are given by:

$$\frac{\partial \Phi_4}{\partial v_i^{\top}} = (4 + 2b_i)v_i^{\top} - 2v_{i+1}^{\top} - 2v_{\cdot}^{\top}{}_{-1} + c_i + \frac{1}{2}(a_{i-1} - a_{i+1}).$$

Thus, we have a system of $n$ equations of the form:

$$(4 + 2b_i)v_i^{\top} - 2v_{i+1}^{\top} - 2v_{i-1}^{\top} = -c_i + \frac{1}{2}(a_{i+1} - a_{i-1}).$$

The continuous forms of the coefficients $a_i$, $b_i$ and $c_i$ are given by the following expressions:

$$A = 2v^{\perp} \left( \mathbf{u}^{\top} \cdot \frac{\partial \mathbf{u}^{\perp}}{\partial s} \right)$$

$$B = \left(\frac{\partial \mathbf{u}^{\top}}{\partial s} \cdot \frac{\partial \mathbf{u}^{\top}}{\partial s}\right) = \left|\frac{\partial \mathbf{u}^{\top}}{\partial s}\right|^2 = \kappa^2$$

$$C = -2\frac{\partial v^{\perp}}{\partial s}\left(\mathbf{u}^{\top} \cdot \frac{\partial \mathbf{u}^{\perp}}{\partial s}\right)$$

where $\kappa$ is the curvature of the curve.

Let the above system of linear equations be denoted by the matrix equation $Bv^{\top} = C$. If $b_i = 0$ for all $i$, then the matrix $B$ is singular. This situation arises when the curve is an extended straight line, for which there is no unique velocity field solution. For all other cases, the matrix $B$ is nonsingular, giving rise to a unique velocity field solution.

There are many available methods for solving systems of linear equations. In the first part of this appendix, we described the methods of *Jacobi* and *Gauss-Seidel* relaxation. For the Jacobi relaxation scheme, the iterative equation for this problem is given by:

$$v_i^{\top(k+1)} = \left(\frac{1}{4 + 2b_i}\right)\left(2v_{i-1}^{\top(k)} + 2v_{i+1}^{\top(k)} - c_i + \frac{1}{2}(a_{i+1} - a_{i-1})\right)$$

for $1 \leq i \leq n$. For Gauss-Seidel relaxation, the iterative equation is given by:

$$v_i^{\top(k+1)} = \left(\frac{1}{4 + 2b_i}\right)\left(2v_{i-1}^{\top(k+1)} + 2v_{i+1}^{\top(k)} - c_i + \frac{1}{2}(a_{i+1} - a_{i-1})\right)$$

for $1 \leq i \leq n$. The matrix $B$ is positive definite, which implies that both the Jacobi and Gauss-Seidel methods will converge.

### A.2 Computation of the *x* and *y* Components of Velocity

A disadvantage of the above formulation of the velocity field algorithm is that in addition to the measurements of $\mathbf{u}^{\perp}(s)$ and $v^{\perp}(s)$, it requires the explicit measurement of $\frac{\partial v^{\perp}}{\partial s}$, $\frac{\partial \mathbf{u}^{\top}}{\partial s}$ and $\frac{\partial \mathbf{u}^{\perp}}{\partial s}$. These additional measurements introduce considerable error to the velocity field solution, and can be avoided by formulating the algorithm in such a way that the $x$ and $y$ components of velocity are computed explicitly, rather than the tangential component. As we noted in Section 4.1, the continuous functional can be expressed in terms of the $x$ and $y$ components of velocity as follows:

$$\Theta = \int \left[ \left( \frac{\partial \mathbf{V}_x}{\partial s} \right)^2 + \left( \frac{\partial \mathbf{V}_y}{\partial s} \right)^2 \right] ds.$$

If there are $n$ points along a contour, the discrete objective function can then be given by:

$$\Phi_1 = \sum_{i=2}^{n} \left[ (\mathbf{V}_{x_i} - \mathbf{V}_{x_{i-1}})^2 + (\mathbf{V}_{y_i} - \mathbf{V}_{y_{i-1}})^2 \right] + \left[ (\mathbf{V}_{x_1} - \mathbf{V}_{x_n})^2 + (\mathbf{V}_{y_1} - \mathbf{V}_{y_n})^2 \right].$$

This expression assumes that the contour is closed, and the points are evenly spaced along the contour. For the case of an open contour, the discrete differences at the endpoints change. If points on the contour are not evenly spaced, the discrete differences can be divided by the distances between adjacent points on the contour.

For the case in which the constraints from the image are to be satisfied exactly, these constraints give rise to a set of equations of the following form:

$$\mathbf{V} \cdot \mathbf{u}^{\perp} - v^{\perp} = 0.$$

Letting $\mathbf{V} = (\mathbf{V}_x, \mathbf{V}_y)$ and $\mathbf{u}^{\perp} = (\mathbf{u}_x^{\perp}, \mathbf{u}_y^{\perp})$, the constraints are simple linear constraints:

$$\mathbf{V}_{x_i} \mathbf{u}_{x_i}^{\perp} + \mathbf{V}_{y_i} \mathbf{u}_{y_i}^{\perp} - v_i^{\perp} = 0.$$

This equation can be used to express the $y$ components of velocity in terms of the $x$ components:

$$\mathbf{V}_{y_i} = \frac{v_i^{\perp} - \mathbf{V}_{x_i} \mathbf{u}_{x_i}^{\perp}}{\mathbf{u}_{y_i}^{\perp}}.$$

We noted in Section 4.2 that these equations are only valid for points at which $\mathbf{u}_{y_i}^{\perp} \neq 0$. In this case, the expressions for $\mathbf{V}_{y_i}$ can be substituted directly into the discrete function $\Phi_1$, yielding the following discrete objective function:

$$\Phi_2 = \sum_{i=2}^{n} \left[ (\mathbf{V}_{x_i} - \mathbf{V}_{x_{i-1}})^2 + \left( \frac{v_i^{\perp}}{\mathbf{u}_{y_i}^{\perp}} - \frac{\mathbf{u}_{x_i}^{\perp}}{\mathbf{u}_{y_i}^{\perp}} \mathbf{V}_{x_i} - \frac{v_{i-1}^{\perp}}{\mathbf{u}_{y_{i-1}}^{\perp}} + \frac{\mathbf{u}_{x_{i-1}}^{\perp}}{\mathbf{u}_{y_{i-1}}^{\perp}} \mathbf{V}_{x_{i-1}} \right)^2 \right]$$

$$+ (\mathbf{V}_{x_1} - \mathbf{V}_{x_n})^2 + \left( \frac{v_1^{\perp}}{\mathbf{u}_{y_1}^{\perp}} - \frac{\mathbf{u}_{x_1}^{\perp}}{\mathbf{u}_{y_1}^{\perp}} \mathbf{V}_{x_1} - \frac{v_n^{\perp}}{\mathbf{u}_{y_n}^{\perp}} + \frac{\mathbf{u}_{x_n}^{\perp}}{\mathbf{u}_{y_n}^{\perp}} \mathbf{V}_{x_n} \right)^2.$$

The goal of the computation is to find the set of $x$ components of velocity, $\{\mathbf{V}_{x_1}, \mathbf{V}_{x_2}, ..., \mathbf{V}_{x_n}\}$ that minimizes $\Phi_2$. This set of components satisfies the system of linear equations given by:

$$\frac{\partial \Phi_2}{\partial \mathbf{V}_{x_i}} = 0, \qquad 1 \leq i \leq n.$$

These equations are of the form:

$$A_i \mathbf{V}_{x_i} - B_i \mathbf{V}_{x_{i-1}} - C_i \mathbf{V}_{x_{i+1}} = D_i$$

where the coefficients are given by the following expressions:

$$A_i = 2 + 2\left(\frac{\mathbf{u}_{x_i}^{\perp}}{\mathbf{u}_{y_i}^{\perp}}\right)^2$$

$$B_i = 1 + \left(\frac{\mathbf{u}_{x_i}^{\perp}}{\mathbf{u}_{y_i}^{\perp}}\right)\left(\frac{\mathbf{u}_{x_{i-1}}^{\perp}}{\mathbf{u}_{y_{i-1}}^{\perp}}\right)$$

$$C_i = 1 + \left(\frac{\mathbf{u}_{x_i}^{\perp}}{\mathbf{u}_{y_i}^{\perp}}\right)\left(\frac{\mathbf{u}_{x_{i+1}}^{\perp}}{\mathbf{u}_{y_{i+1}}^{\perp}}\right)$$

$$D_i = \left(\frac{\mathbf{u}_{x_i}^{\perp}}{\mathbf{u}_{y_i}^{\perp}}\right)\left[2\left(\frac{v_i^{\perp}}{\mathbf{u}_{y_i}^{\perp}}\right) - \frac{v_{i-1}^{\perp}}{\mathbf{u}_{y_{i-1}}^{\perp}} - \frac{v_{i+1}^{\perp}}{\mathbf{u}_{y_{i+1}}^{\perp}}\right].$$

If we let $\alpha_i$ denote the angle between the vertical axis and the perpendicular vector $\mathbf{u}^{\perp}(s)$ at a point $(x_i, y_i)$ on the contour, then the coefficients may be rewritten as follows:

$$A_i = 2 + 2\tan^2 \alpha_i$$

$$B_i = 1 + \tan \alpha_i \tan \alpha_{i-1}$$

$$C_i = 1 + \tan \alpha_i \tan \alpha_{i+1}$$

$$D_i = \tan \alpha_i \left[\frac{2v_i^{\perp}}{\sin \alpha_i} - \frac{v_{i-1}^{\perp}}{\sin \alpha_{i-1}} - \frac{v_{i+1}^{\perp}}{\sin \alpha_{i+1}}\right].$$

As before, methods such as Jacobi and Gauss-Seidel relaxation can be used to solve the above system of linear equations. Let $\mathbf{V}_x^{(k+1)}$ denote the new approximation for

$\mathbf{V}_x$. For the Jacobi method, the iterative equation is given by:

$$\mathbf{V}_{x_i}^{(k+1)} = \frac{1}{A_i}\Big[B_i \mathbf{V}_{x_{i-1}}^{(k)} + C_i \mathbf{V}_{x_{i+1}}^{(k)} + D_i\Big].$$

For Gauss-Seidel relaxation, this equation is given by:

$$\mathbf{V}_{x_i}^{(k+1)} = \frac{1}{A_i}\Big[B_i \mathbf{V}_{x_{i-1}}^{(k+1)} + C_i \mathbf{V}_{x_{i+1}}^{(k)} + D_i\Big].$$

Along a straight line, these equations take on a particularly simple form. The iterative equation for the Jacobi method is then given by:

$$\mathbf{V}_{x_i}^{(k+1)} = \frac{1}{2}\mathbf{V}_{x_{i-1}}^{(k)} + \frac{1}{2}\mathbf{V}_{x_{i+1}}^{(k)} + \rho\Big[2v_i^{\perp} - v_{i-1}^{\perp} - v_{i+1}^{\perp}\Big]$$

where $\rho = 2\cos\alpha(1 + \tan^2\alpha)$ is a constant factor that depends only on the orientation of the line.

The above analysis is valid only for the case in which $\mathbf{u}_{y_i}^{\perp} \neq 0$, for all points $(x_i, y_i)$. It may be the case, however, that for some points, $\mathbf{u}_{y_i}^{\perp} = 0$. We noted in Section 4.2 that at these points, the value of $\mathbf{V}_{x_i}$ is known, and can be substituted directly into the discrete objective function, while the component $\mathbf{V}_{y_i}$ is unknown, and therefore can be left as a parameter. This still results in a system of $n$ linear equations in $n$ unknowns. The $x$ and $y$ components of velocity are related to the tangential and perpendicular components by a local rotation, and are therefore related by a linear combination. The decomposition of velocity into tangential and perpendicular components led to a system of linear equations for which a unique solution was guaranteed, except in the case of an extended straight line. This is also the case for the decomposition of velocity into $x$ and $y$ components.

### B. The Method of Conjugate Gradients

In Section 4.3, it was shown that the velocity field computation, in which the constraints imposed by the measurements of $v^{\perp}(s)$ are only approximately satisfied, can be formulated as an unconstrained optimization problem, and the conjugate gradient algorithm can be used to obtain a solution. This computation finds the $x$ and $y$ components of velocity for each of the $n$ points on a contour. Each possible solution can be considered as a point in a $2n$-dimensional vector space, with each dimension representing the $x$ or $y$ component of velocity at one point on the contour. The $2n$-dimensional vector space can be embedded in a $(2n+1)$-dimensional vector space, where the final dimension corresponds to the total variation of velocity along the contour. By associating a measure of variation with each possible velocity field, we can construct a *hypersurface*, called the *objective surface*, in the $(2n+1)$-dimensional space. The goal of the velocity field computation is to find the "lowest" point on this surface; this point corresponds to the velocity field of least variation.

The conjugate gradient algorithm is an iterative descent algorithm; a sequence of approximations $\mathbf{V}^{(1)}, \mathbf{V}^{(2)}, \ldots$ to the exact solution $\mathbf{V}$ is computed, given an initial approximation $\mathbf{V}^{(0)}$, and this sequence has the property that each new approximation decreases the value of the objective function. The basic steps for a descent method are as follows:

(1) Start at an initial point $\mathbf{V}^{(0)}$, which is usually $\mathbf{V}^{(0)} = 0$.

(2) According to a fixed rule, determine a direction of movement along the objective surface that will reduce the value of the objective function.

(3) Move in this direction to a (relative) minimum of the objective function.

(4) If the final solution has not yet been reached, return to step 2.

One of the most common descent methods is that of *steepest descent*, in which the direction of movement in step 2 above is given by the negative gradient of the objective function, evaluated at the present point. The steepest descent method has the disadvantage of relatively slow convergence. A much improved rate of convergence can be obtained with the method of conjugate gradients.

The velocity field of least variation is given by the solution of a system of linear equations. Let this system be denoted in matrix form by:

$$\mathbf{Qx} = \mathbf{b}$$

where $\mathbf{Q}$ is an $n \times n$ symmetric positive definite matrix. The unique solution to this equation is equivalent to the unique solution of the quadratic problem:

$$\begin{aligned} &\text{minimize} && \frac{1}{2}\mathbf{x}^T\mathbf{Qx} - \mathbf{b}^T\mathbf{x} \\ &\text{subject to} && \mathbf{x} \in \Omega \subset E^n. \end{aligned}$$

$E^n$ denotes the $n$-dimensional Euclidean space. Conjugate direction methods were invented largely for the analysis of quadratic problems. We present a brief development of these methods, and the conjugate gradient algorithm. This development is taken from Luenberger (1973), where further details may be found.

The basic idea behind conjugate direction methods is to optimize the direction in which each new step is taken in the algorithm. To show how this optimization is achieved, we begin by defining the notion of **Q**-orthogonality:

**Definition.** *Given a symmetric matrix* $\mathbf{Q}$, *two vectors* $\mathbf{d}_1$ *and* $\mathbf{d}_2$ *are said to be* $\mathbf{Q}$*-orthogonal, or conjugate, with respect to* $\mathbf{Q}$ *if* $\mathbf{d}_1^\top \mathbf{Q} \mathbf{d}_2 = 0$.

The following can then be shown:

**Proposition.** *If* $\mathbf{Q}$ *is positive definite and the set of nonzero vectors* $\mathbf{d}_0$, $\mathbf{d}_1, \ldots, \mathbf{d}_k$ *are* $\mathbf{Q}$*-orthogonal, then these vectors are linearly independent.*

Given the positive definite matrix $\mathbf{Q}$, and a set of non-zero $\mathbf{Q}$-orthogonal vectors, $\mathbf{d}_0, \mathbf{d}_1, \ldots \mathbf{d}_{n-1}$, their linear independence allows us to express the solution $\mathbf{x}'$ in terms of these vectors:

$$\mathbf{x}' = \alpha_0 \mathbf{d}_0 + \cdots + \alpha_{n-1} \mathbf{d}_{n-1}$$

for some set of scalars $\alpha_i$. The orthogonality of the vectors implies that multiplying by $\mathbf{Q}$ and taking the scalar product with $\mathbf{d}_i$ causes all the terms, except the $i^{th}$, to vanish, yielding:

$$\alpha_i = \frac{\mathbf{d}_i^\top \mathbf{Q} \mathbf{x}'}{\mathbf{d}_i^\top \mathbf{Q} \mathbf{d}_i} = \frac{\mathbf{d}_i^\top \mathbf{b}}{\mathbf{d}_i^\top \mathbf{Q} \mathbf{d}_i}.$$

The solution is then given by the following:

$$\mathbf{x}' = \sum_{i=0}^{n-1} \frac{\mathbf{d}_i^\top \mathbf{b}}{\mathbf{d}_i^\top \mathbf{Q} \mathbf{d}_i} \mathbf{d}_i.$$

The expansion for the solution $\mathbf{x}'$ can be considered as the result of an iterative process of $n$ steps, where at the $i^{th}$ step, the term $\alpha_i \mathbf{d}_i$ is added. In this way, the following result can be proven (see, for example, Luenberger 1973):

**Conjugate Direction Theorem:** *Let* $\{\mathbf{d}_i\}_{i=0}^{n-1}$ *be a set of non-zero* **Q**-*orthogonal vectors. For any* $\mathbf{x}_0 \in E^n$ *the sequence* $\{\mathbf{x}_k\}$ *generated according to*

$$\mathbf{x}_{k+1} = \mathbf{x}_k + \alpha_k \mathbf{d}_k, \qquad k \leq 0$$

*with*

$$\alpha_k = -\frac{\mathbf{g}_k^\top \mathbf{d}_k}{\mathbf{d}_k^\top \mathbf{Q} \mathbf{d}_k}$$

*and*

$$\mathbf{g}_k = \mathbf{Q}\mathbf{x}_k - \mathbf{b}$$

*converges to the unique solution,* $\mathbf{x}'$, *of* $\mathbf{Q}\mathbf{x} = \mathbf{b}$ *after* $n$ *steps.*

The conjugate gradient algorithm provides one method for generating the sequence of **Q**-orthogonal direction vectors; the successive direction vectors are selected as a conjugate version of the successive gradients obtained as the method proceeds. Hence, at each step, the next direction vector is determined, based on the current state of the objective function and its gradient vector. The algorithm, which is guaranteed to converge in at most $n$ steps, is outlined below:

**The Conjugate Gradient Algorithm**

Starting at any point $\mathbf{x}_0 \in E^n$ define $\mathbf{d}_0 = -\mathbf{g}_0 = \mathbf{b} - \mathbf{Q}\mathbf{x}_0$ and

$$\mathbf{x}_{k+1} = \mathbf{x}_k + \alpha_k \mathbf{d}_k$$

$$\alpha_k = -\frac{\mathbf{g}_k^T \mathbf{d}_k}{\mathbf{d}_k^T \mathbf{Q} \mathbf{d}_k}$$

$$\mathbf{d}_{k+1} = -\mathbf{g}_{k+1} + \beta_k \mathbf{d}_k$$

$$\beta_k = \frac{\mathbf{g}_{k+1}^T \mathbf{Q} \mathbf{d}_k}{\mathbf{d}_k^T \mathbf{Q} \mathbf{d}_k}$$

where $\mathbf{g}_k = \mathbf{Q}\mathbf{x}_k - \mathbf{b}$.

For the velocity field computation, we showed in Section 4.3 that the system of $2n$ linear equations to be solved is given by:

$$\frac{\partial \Phi_3}{\partial \mathbf{V}_{x_i}} = 0, \qquad \frac{\partial \Phi_3}{\partial \mathbf{V}_{y_i}} = 0, \qquad 1 \leq i \leq n$$

where $\Phi_3$ is given by:

$$\Phi_3 = \sum_{i=2}^{n} \left[ (\mathbf{V}_{x_i} - \mathbf{V}_{x_{i-1}})^2 + (\mathbf{V}_{y_i} - \mathbf{V}_{y_{i-1}})^2 \right] + \left[ (\mathbf{V}_{x_1} - \mathbf{V}_{x_n})^2 + (\mathbf{V}_{y_1} - \mathbf{V}_{y_n})^2 \right]$$

$$+\beta \sum_{i=1}^{n} \left[ \mathbf{V}_{x_i} \mathbf{u}_{x_i}^{\perp} + \mathbf{V}_{y_i} \mathbf{u}_{y_i}^{\perp} - v_i^{\perp} \right]^2 .$$

From the partial derivatives with respect to the $x$ components of velocity, we have the equations:

$$(4 + 2\beta(\mathbf{u}_{x_i}^{\perp})^2)\mathbf{V}_{x_i} - 2\mathbf{V}_{x_{i+1}} - 2\mathbf{V}_{x_{i-1}} + 2\beta \mathbf{u}_{x_i}^{\perp} \mathbf{u}_{y_i}^{\perp} \mathbf{V}_{y_i} = 2\beta v_i^{\perp} \mathbf{u}_{x_i}^{\perp}, \qquad 1 \leq i \leq n.$$

From the partial derivatives with respect to the $y$ components of velocity, we have:

$$(4 + 2\beta(\mathbf{u}_{y_i}^{\perp})^2)\mathbf{V}_{y_i} - 2\mathbf{V}_{y_{i+1}} - 2\mathbf{V}_{y_{i-1}} + 2\beta \mathbf{u}_{x_i}^{\perp} \mathbf{u}_{y_i}^{\perp} \mathbf{V}_{x_i} = 2\beta v_i^{\perp} \mathbf{u}_{y_i}^{\perp}, \qquad 1 \leq i \leq n.$$

The coefficients on the left hand side of the equations form the symmetric positive definite matrix **Q**, and the values on the right hand side form the column vector **b**. It should be noted that for the case of an extended straight line, **Q** is singular, so that a unique solution does not exist. The initial velocity field consists of the normal velocity vectors; hence:

$$\mathbf{x}_0 = (v_1^{\perp} \mathbf{u}_{x_1}^{\perp},\ v_2^{\perp} \mathbf{u}_{x_2}^{\perp},\ \ldots,\ v_n^{\perp} \mathbf{u}_{x_n}^{\perp},\ v_1^{\perp} \mathbf{u}_{y_1}^{\perp},\ v_2^{\perp} \mathbf{u}_{y_2}^{\perp},\ \ldots,\ v_n^{\perp} \mathbf{u}_{y_n}^{\perp}).$$

The conjugate gradient algorithm is then applied straightforwardly.

# Appendix V

In this appendix, we present details of the perceptual experiments described in Chapters 6 and 7. The appendix is divided into two parts, which address experiments relevant to the velocity field computation, and detection of motion discontinuities, respectively.

## The Velocity Field Computation

In this section, we present details of the following demonstrations:

1. The reversed phi phenomenon embedded in moving bars (Figure 6.6a).
2. Vertical bars moving to bars with trapezoid-shaped edges (Figure 6.6b).
3. Vertical bars moving to angle brackets (Figure 6.7a).
4. Translating rigid figures (Figure 6.8).
5. Rotating smooth curves (ie, the circular helix (Figure 4.8) and the circle wrapped around a cylinder (Figures 4.10 and 4.13)).
6. The accelerating and expanding circle (Section 6.2.2).

In the discussion that follows, we refer to these demonstrations by number.

Two systems were used for the creation and display of the above stimulus patterns. Some patterns (demonstrations 1 and 5) were generated on a LISP machine and displayed on a high resolution ($768 \times 1024$ points, with 0.026 cm/point), 60 Hz, non-interlaced, black and white video monitor with P4 phosphor. Uniform grey, required for demonstration 1, was created by turning on every other point in the display, in a checkerboard pattern. From the viewing distances used, the individual points were not distinguishable. Mean luminance of the display was not carefully controlled for these demonstrations. Presentation time and interstimulus interval (ISI) were specified in 60ths of a second. Other stimuli (demonstrations 2, 3, 4 and 6) were presented on a Vector General display unit with fast decaying P4 phosphor. Combined with DEC PDP-11/45 and PDP-10 computers, this system provides for the creation, storage and display of line drawings, with control of presentation time and ISI in milliseconds. For all demonstrations, the room was dark, the display was viewed monocularly, and fixation was maintained on a small black dot placed at the center of the display.

Most of the above motion displays were viewed by two observers. The author served as one observer, and the second varied. All observers were experienced in viewing apparent movement stimuli under conditions of steady fixation. The moving pattern was presented for at least a few seconds, and the observers reported the direction of movement of different parts of the pattern. In general, the perception of direction of motion was immediate, and did not change during the course of the presentation. Exceptions were noted in the text.

Details of the spatial and temporal parameters for the first four demonstrations were described in the text. For the smooth curves in rotation (ie, the circular helix, and the circle wrapped around a cylinder), a frame was created every $5°$ in the curve's rotation. An entire revolution thus consisted of 72 frames. The curves filled an area of the display whose size was roughly $10° \times 10°$ of visual arc. Displacements between frames were within the short range limits (ie, less than 15' of visual arc), and presentation rate was roughly 30 frames/second. The movies were presented for several seconds, while fixation was maintained at a central point on the display. The observer then provided a qualitative description of the perceived movement and three-dimensional structure of the curve.

**The Detection of Motion Discontinuities**

In this section, we discuss the details of two experiments relating to the detection of motion discontinuities, presented in Chapter 7:

1. The boundary localization experiments (Figure 7.1a and Section 7.2).

2. The size versus displacement experiments (Figure 7.1b and Section 7.4).

The random dot patterns were generated on a LISP machine and displayed on the high resolution black and white video monitor described above. The patterns consisted of $160 \times 160$ dots, each of size $3' \times 3'$ of visual arc. Thus, the entire pattern subtended an area of size $8° \times 8°$ of visual arc. The subject viewed the display monocularly, from a distance of 30 cm, in a darkened room. Fixation was maintained on a mark at the center of the display.

For the first experiments, velocity changed either in direction or in magnitude across a vertical boundary. Six angular differences (18.4°, 26.6°, 36.9°, 45°, 53.1° and 61.4°) and three speed differences (2:1, 3:1 and 4:1) were used. The absolute directions of velocity varied over the four quadrants, yielding a total of

24 combinations of velocities for which there was a change in direction, and 12 combinations for which there was a change in magnitude.

A typical experimental session consisted of 100 to 200 trials. The combination of velocities of the two fields of dots, horizontal position of the boundary, and vertical position of the notch were varied randomly from one trial to the next. For each trial, one frame first appeared on the display. On the subject's signal, four additional frames appeared, in rapid succession. ISI was a few msec, and total duration of movement was roughly 80 — 90 msec. The subject then indicated whether the notch appeared above or below the point of fixation. A sufficient number of sessions was run to obtain at least 50 trials for each combination of parameters.

In the second set of experiments, a small square patch was displaced in the direction opposite the direction of motion of the background. Four displacements of the patch and background were used (3', 6', 9' and 12' of visual arc). For each displacement size, four different patch sizes were tested. The points in Figure 7.5 indicate the particular combinations of patch and displacement sizes. Absolute direction of movement of the background varied over four directions (right, left, up and down), giving a total of 64 combinations of patch size, displacement, and direction of motion.

Again, a typical experimental session consisted of 100 to 200 trials. The combination of motion parameters, and quadrant in which the patch appeared, were varied randomly from one trial to the next. For each trial, one frame first appeared on the display. On the subject's signal, four additional frames appeared, in rapid succession. ISI was a few msec, and total duration of movement was roughly 80 — 90 msec. The subject then indicated which quadrant contained the differentially moving patch. A sufficient number of sessions was run to obtain at least 50 trials for each combination of patch size and displacement.

## References


**Adelson, E. H. and Movshon, J. A.** (1982) "Phenomenal Coherence of Moving Visual Patterns," *Nature 300*, 523-525.

**Adelson, E. H. and Movshon, J. A.** (1983) "The Perception of Coherent Motion in Two-dimensional Patterns," *Proceedings of the ACM Interdisciplinary Workshop on Motion: Representation and Perception*, Toronto, Canada, 11-14.

**Aggarwal, J. K. and Duda, R. O.** (1975) "Computer Analysis of Moving Polygonal Images," *IEEE Transactions on Computers, c-24*, 966-976.

**Albrecht, D. G. and Hamilton, D. B.** (1982) "Striate Cortex of Monkey and Cat: Contrast Response Function," *Journal of Neurophysiology 48*, 217-237.

**Albright, T. D. and Gross, C. G.** (1982) "Direction and Orientation Selectivity of Neurons in Area MT of Macaques," *Neuroscience Abstracts*, 811.

**Alpern, M.** (1954) *Americal Medical Association Archives of Opthalmology 51*, 369-374.

**Andrews, D. P., Mackay, D. M. and Wilson, J. P.** (1963) "Comment on an Article entitled, "Apparent Relative Movement of 'Unsharp' and 'Sharp' Visual Patterns," by F. J. Verheijen," *Nature 199*, 161.

**Anstis, S. M.** (1967) "Visual Adaptation to Gradual Change of Intensity," *Science 155*, 710-712.

**Anstis, S. M.** (1970) "Phi Movement as a Subtraction Process," *Vision Research 10*, 1411-1430.

**Anstis, S. M.** (1980) "The Perception of Apparent Motion," *Philosophical Transactions of the Royal Society of London B. 290*, 153-168.

**Anstis, S. M.** (1983) "Visual Coding of Position and Motion," in: *Physical and Biological Processing of Images.* ed. by O. J. Braddick and A. C. Sleigh, Springer-Verlag, Berlin.

**Anstis, S. M. and Rogers, B. J.** (1975) "Illusory Reversal of Visual Depth and Movement During Changes of Contrast," *Vision Research 15*, 957-961.

**Attneave, F. and Block, G.** (1973) "Apparent Motion in Tridimensional Space," *Perception and Psychophysics 13*, 301-307.

**Baker, C. L. and Braddick, O. J.** (1982a) "Does Segregation of Differently Moving Areas Depend on Relative or Absolute Displacement," *Vision Research 7*, 851-856.

**Baker, C. L. and Braddick, O. J.** (1982b) "The Basis of Area and Dot Number Effects in Random Dot Motion Perception," *Vision Research 10*, 1253-1260.

**Baker, J. F., Petersen, S. E., Newsome, W. T. and Allman, J. M.** (1981) "Visual Response Properties of Neurons in Four Extrastriate Visual Areas of the Owl Monkey (Aotus Trivirgatus): a Quantitative Comparison of Medial, Dorsomedial, Dorsolateral, and Middle Temporal Areas," *Journal of Neurophysiology 45*, 397-416.

**Barlow, H. B. and Levick, R. W.** (1965) "The Mechanism of Directional Selectivity in the Rabbit's Retina," *Journal of Physiology (London) 173*, 477-504.

**Barnard, S. T. and Thompson, W. B.** (1980) "Disparity Analysis of Images," *IEEE Transactions on Pattern Analysis and Machine Intelligence 2*, 333-340.

**Batali, J. and Ullman, S.** (1979) "Motion Detection and Analysis," *Proceedings of the ARPA Image Understanding Workshop*, ed. by L. S. Baumann, Science Applications Inc., Arlington, VA, 69-75.

**Baylor, D. A., Hodgkin, A. L. and Lamb, T. D.** (1974) "Reconstruction of the Electrical Responses of Turtle Cones to Flashes and Steps of Light," *Journal of Physiology (London) 242*, 759-791.

**Bell, H. H. and Lappin, J. S.** (1973) "Sufficient Conditions for the Discrimination of Motion," *Perception and Psychophysics 11*, 263-268.

**Bell, H. H. and Lappin, J. S.** (1979) "The Detection of Rotation in Random-dot Patterns," *Perception and Psychophysics 26*, 415-417.

**Benevento, L. A. and Rezak, M.** (1976) "The Cortical Projections of the Inferior Pulvinar and Adjacent Lateral Pulvinar in the Rhesus Monkey *(Macaca Mulatta)*: An Autoradiographic Study," *Brain Research 108*, 1-24.

**Bishop, P. O., Coombs, J. S. and Henry, G. H.** (1971a) "Responses to Visual Contours: Spatio-temporal Aspects of Excitation in the Receptive Fields of Simple Striate Neurons," *Journal of Physiology (London) 219*, 625-657.

**Bishop, P. O., Coombs, J. S. and Henry, G. H.** (1971b) "Interaction Effects of Visual Contours on the Discharge Frequency of Simple Striate Neurons," *Journal of Physiology (London) 219*, 659-687.

**Blakemore, C. and Campbell, F. W.** (1969) "On the Existence of Neurones in the Human Visual System Selectively Sensitive to the Orientation and Size of Retinal Images," *Journal of Physiology (London) 203*, 237-160.

**Bobick, A.** (1983) "A Hybrid Approach to Structure-from-motion," *Proceedings of the ACM Interdisciplinary Workshop on Motion: Representation and Perception*, Toronto, Canada, 91-109.

**Boycott, B. B. and Wassle, H.** (1974) "The Morphological Types of Ganglion Cells of the Domestic Cat's Retina," *Journal of Physiology (London) 240*, 397-419.

**Braddick, O. J.** (1973) "The Masking of Apparent Motion in Random-dot Patterns," *Vision Research 13*, 355-369.

**Braddick, O. J.** (1974) "A Short-range Process in Apparent Motion," *Vision Research 14*, 519-527.

**Braddick, O. J.** (1980) "Low-level and High-level Processes in Apparent Motion," *Philosophical Transactions of the Royal Society of London B. 290*, 137-151.

**Brady, J. M.** (1982) "Computational Approaches to Image Understanding," *Computing Surveys 14*, 3-71.

**Brandt, A.** (1977a) "Multi-level Adaptive Solutions to Boundary Value Problems," *Mathematical Computing 31*, 333-390.

**Brandt, A.** (1977b) "Multi-level Adaptive Techniques (MLAT) for Partial Differential Equations," *Mathematical Software III*, ed. by J. R. Rice, Academic Press, New York, 277-318.

**Breitmeyer, B. G. and Ganz, L.** (1977) "Temporal Studies with Flashed Gratings: Inferences About Human Transient and Sustained Channels," *Vision Research 17*, 861-866.

**Bruss, A. and Horn, B. K. P.** (1983) "Passive Navigation," *Computer Vision, Graphics, and Image Processing 21*, 3-20.

**Bulthoff, H. and Gotz, K. G.** (1979) "Analogous Motion Illusion in Man and Fly," *Nature 278*, 636-638.

**Burr, D. C.** (1981) "Temporal Summation of Moving Images by the Human Visual System," *Proceedings of the Royal Society of London B. 211*, 321-339.

**Burr, D. C., Morrone, C. and Maffei, L.** (1981) "Intra-cortical Inhibition Prevents Simple Cells from Responding to Textured Visual Patterns," *Experimental Brain Research 43*, 455-458.

**Burr, D. C. and Ross, J.** (1982) "Contrast Sensitivity at High Velocities," *Vision Research 22*, 479-484.

**Burt, P. and Sperling, G.** (1981) "Time, Distance, and Feature Trade-offs in Visual Apparent Motion," *Psychological Review 88*, 171-195.

**Campbell, F. W. and Robson, J. G.** (1968) "Applications of Fourier Analysis to the Visibility of Gratings," *Journal of Physiology (London) 197*, 551-556.

**Chang, J. J. and Julesz, B.** (1983) "Displacement Limits, Directional Anisotropy and Direction Versus Form Discrimination in Random-dot Cinematograms," *Vision Research 23*, 639-646.

**Chow, W. K. and Aggarwal, J. K.** (1977) "Computer Analysis of Planar Curvilinear Moving Images," *IEEE Transactions on Computers c-26*, 179-185.

**Clatworthy, J. L. and Frisby, J. P.** (1973) "Real and Apparent Movement: Evidence for Unitary Mechanism," *Perception 2*, 161-164.

**Cleland, B. G., Dubin, M. W. and Levick, W. R.** (1971) "Sustained and Transient Neurons in the Cat's Retina and LGN," *Journal of Physiology (London) 217*, 473-496.

**Cleland, B. G. and Levick, W. R.** (1974) "Properties of Rarely Encountered Types of Ganglion Cells in the Cat's Retina and Overall Classification," *Journal of Physiology (London) 240*, 457-492.

**Cleland, B. G., Levick, W. R. and Sanderson, K. J.** (1973) "Properties of Sustained and Transient Ganglion Cells in the Cat Retina," *Journal of Physiology (London) 228*, 649-680.

**Clocksin, W. F.** (1980) "Perception of Surface Slant and Edge Labels from Optical Flow: a Computational Approach," *Perception 9*, 253-269.

**Corbin, H. H.** (1942) "The Perception of Grouping and Apparent Motion in Visual Space," *Archives of Psychology 273.*

**Cowan, J. D.** (1977) "Some Remarks on Channel Bandwidths for Visual Contrast Detection," *Neurosciences Research Program Bulletin 15*, 492-517.

**Dahlquist, G. and Bjorck, A.** (1974) *Numerical Methods*, Prentice-Hall, Englewood Cliffs, New Jersey.

**Davis, L.** (1975) "A Survey of Edge Detection Techniques," *Computer Graphics and Image Processing 4*, 248-270.

**Davis, L., Wu, Z. and Sun, H.** (1982) "Contour-based Motion Estimation," *Proceedings of the ARPA Image Understanding Workshop*, ed. by L. S. Baumann, Science Applications Inc., Arlington, VA, 124-131.

**Dawis, S. M.** (1978) "A Model for Light Adaptation: Producing Weber's Law With Bleaching-type Kinetics," *Biological Cybernetics 30*, 187-193.

**DeMonasterio, F. M.** (1978a) "Properties of Concentrically Organized X and Y Ganglion Cells of Macaque Retina," *Journal of Neurophysiology 41*, 1394-1417.

**DeMonasterio, F. M.** (1978b) "Center and Surround Mechanisms of Opponent Color X and Y Ganglion Cells of the Retina of Macaques," *Journal of Neurophysiology 41*, 1418-1434.

**DeMonasterio, F. M.** (1978c) "Properties of Ganglion Cells With Atypical Receptive-Field Organization in Retina of Macaques," *Journal of Neurophysiology 41*, 1435-1449.

**DeMonasterio, F. M. and Gouras, P.** (1975) "Functional Properties of Ganglion Cells of the Rhesus Monkey Retina," *Journal of Physiology (London) 251*, 167-195.

**DeValois, R. L., Albrecht, D. G. and Thorell, L. G.** (1982) "Spatial Frequency Selectivity of Cells in Macaque Visual Cortex," *Vision Research 22*, 545-559.

**DeValois, R. L., Yund, E. W. and Hepler, N.** (1982) "The Orientation and Direction Selectivity of Cells in Macaque Visual Cortex," *Vision Research 22*, 531-544.

**Diener, H. C., Wist, E. R., Dichgans, J. and Brandt, TH.** (1976) "The Spatial Frequency Effect on Perceived Velocity," *Vision Research 16*, 169-176.

**Doorn, A. J. Van and Koenderink, J. J.** (1982) "Visibility of Movement Gradients," *Biological Cybernetics 44*, 167-175.

**Doorn, A. J. Van and Koenderink, J. J.** (1983) "Detectability of Velocity Gradients in Moving Random-Dot Patterns," *Vision Research 23*, 799-804.

**Dow, B. M.** (1974) "Functional Classes of Cells and Their Laminar Distribution in Monkey Visual Cortex," *Journal of Neurophysiology 37*, 927-946.

**Dreher, B., Fukada, Y. and Rodieck, R. W.** (1976) "Identification, Classification and Anatomical Segregation of Cells with X-like and Y-like Properties in the Lateral Geniculate Nucleus of Old World Primates," *Journal of Physiology (London) 258*, 433-452.

**Dreher, B. and Sanderson, K. J.** (1973) "Receptive Field Analysis: Responses to Moving Visual Contours by Single Lateral Geniculate Neurons in the Cat," *Journal of Physiology (London) 234*, 95-118.

**Enroth-Cugell, C. and Robson, J. G.** (1966) "The Contrast Sensitivity of Retinal Ganglion Cells of the Cat," *Journal of Physiology (London) 187*, 517-522.

**Exner, S.** (1875) "Ueber das Sehen von Bewegungen und die Theorie des zusammengesetzen Auges," *Sitzungsberichte Akademie Wissenschaft Wien 72*, 156-160.

**Fahle, M. and Poggio, T.** (1981) "Visual Hyperacuity: Spatiotemporal Interpolation in Human Vision," *Proceedings of the Royal Society of London B. 213*, 415-477.

**Fennema, C. I. and Thompson, W. B.** (1979) "Velocity Determination in Scenes Containing Several Moving Objects," *Computer Graphics and Image Processing 9*, 301-315.

**Fukada, Y. and Stone, J.** (1974) "Retinal Distribution and Central Projections of Y-, X-, and W-cells of the Cat's Retina," *Journal of Neurophysiology 37*, 749-772.

**Gattas, R. and Gross, C. G.** (1979) "A Visuotopically Organized Area in the Posterior Superior Temporal Sulcus of the Macaque," *Association for Research in Vision and Ophthalmology, Abstracts*, 184.

**Gattas, R. and Gross, C. G.** (1981) "Visual Topography of Striate Projection Zone (MT) in Posterior Superior Temporal Sulcus of the Macaque," *Journal of Neurophysiology 46*, 621-638.

**Gibson, J. J.** (1950) *The Perception of the Visual World*, Houghton Mifflin, Boston.

**Gladwell, I. and Wait, R., editors** (1979) *A Survey of Numerical Methods for Partial Differential Equations*, Clarendon Press, Oxford.

**Goodwin, A. W., Henry, G. H. and Bishop, P. O.** (1975) "Direction Selectivity of Simple Striate Cells: Properties and Mechanism," *Journal of Neurophysiology 38*, 1500-1523.

**Graham, C. H.** (1965) "Perception of Movement," in: *Vision and Visual Perception*, ed. by C. H. Graham, John Wiley & Sons, New York.

**Graham, N.** (1977) "Visual Detection of Aperiodic Spatial Stimuli by Probability Summation Amoung Narrow Band Channels," *Vision Research 17*. 637-652.

**Graham, N. and Nachmias, J.** (1971) "Detection of Grating Patterns Containing Two Spatial Frequencies: A Comparison of Single-channel and Multiple-channel Models," *Vision Research 11*, 251-259.

**Green, M. and von Grunau, M.** (1983) "Real and Apparent Motion: One Mechanism or Two?" *Proceedings of the ACM Interdisciplinary Workshop on Motion: Representation and Perception*, Toronto, Canada, 17-22.

**Grimson, W. E. L.** (1981) *From Images to Surfaces. A Computational Study of the Human Early Visual System*, MIT Press, Cambridge and London.

**Hadani, I., Gur, M. and Meiri, A. Z.** (1981) "Detection of Differential Displacements of Random Dot Patterns at Different Dot Densities," *Vision Research 21*, 1193-1195.

**Hartenberg, R. S. and Denavit, J.** (1964) *Kinematic Synthesis of Linkages*, McGraw-Hill, New York.

**Hassenstein, B. and Reichardt, W.** (1956) "Systemtheoretische Analyse der Zeit-, Reihenfolgen- und Vorzeichenauswertung bei der Bewgungs-perzeption der Russelkafers," *Chlorophanus. Z. Naturf. 11b*, 513-524.

**Hildreth, E. C.** (1982) "The Integration of Motion Information Along Contours," *Proceedings of the IEEE Workshop on Computer Vision: Representation and Control*, Rindge, NH, August, 83-91.

**Hildreth, E. C.** (1983) "The Detection of Intensity Changes by Computer and Biological Vision Systems," *Computer Vision, Graphics, and Image Processing 22*, 1-27.

**Hochstein, S. and Shapley, R. M.** (1976) "Quantitative Analysis of Retinal Ganglion Cell Classifications," *Journal of Physiology (London) 262*, 237-264.

**Hoffman, D. D. and Flinchbaugh, B. E.** (1982) "The Interpretation of Biological Motion," *Biological Cybernetics 42*, 195-204.

**Hoffman, D. D. and Richards, W. A.** (1982) "Representing Plane Curves for Recognition," *MIT Artificial Intelligence Laboratory Memo 630.*

**Hoffman, K.-P., Stone, J. and Sherman, S. M.** (1972) "Relay of Receptive Field Properties in Dorsal Lateral Geniculate Nucleus of the Cat," *Journal of Neurophysiology 35*, 518-531.

**Holland, H. C.** (1965) *The Spiral After-Effect*, Pergamon Press, Oxford.

**Holvand, C. I. H.** (1935) "Apparent Movement," *Psychological Bulletin 32*, 755-778.

**Horn, B. K. P. and Schunck, B. G.** (1981) "Determining Optical Flow," *Artificial Intelligence 17*, 185-203.

**Hubel, D. H. and Wiesel, T. N.** (1961) "Integrative Action in the Cat's Lateral Geniculate Body," *Journal of Physiology (London) 155*, 385-398.

**Hubel, D. H. and Wiesel, T. N.** (1962) "Receptive Fields, Binocular Interaction, and Functional Architecture in the Cat's Visual Cortex," *Journal of Physiology (London) 160*, 106-154.

**Hubel, D. H. and Wiesel, T. N.** (1968) "Receptive Fields and Functional Architecture of Monkey Striate Cortex," *Journal of Physiology (London) 195*, 215-243.

**Hubel, D. H. and Wiesel, T. N.** (1972) "Laminar and Columnar Distribution of Geniculo-Cortical Fibers in the Macaque Monkey," *Journal of Comparative Neurology 146*, 421-450.

**Ikeda, H. and Wright, M. J.** (1972) "Receptive Field Organization of 'Sustained' and 'Transient' Retinal Ganglion Cells Which Subserve Different Functional Roles," *Journal of Physiology (London) 227*, 769-800.

**Ikeda, H. and Wright, M. J.** (1975) "Spatial and Temporal Properties of "Sustained" and "Transient" Neurons in Area 17 of the Cat's Visual Cortex," *Experimental Brain Research 22*, 363-383.

**Jain, R., Martin, W. N. and Aggarwal, J. K.** (1979) "Segmentation Through the Detection of Changes Due to Motion," *Computer Graphics and Image Processing 11*, 13-34.

**Johansson, G.** (1973) "Visual Perception of Biological Motion and a Model for its Analysis," *Perception and Psychophysics 14*, 201-211.

**Johansson, G.** (1975) "Visual Motion Perception," *Scientific American 232*, 76-88.

**Julesz, B.** (1971) *Foundations of Cyclopean Perception*, The University of Chicago Press, Chicago.

**Kahneman, D. and Wolman, R.** (1970) "Stroboscopic Motion: Effects of Duration and Interval," *Perception and Psychophysics 8*, 161-164.

**Kolers, P. A.** (1972) *Aspects of Motion Perception*, Pergamon Press, New York.

**Korte, A.** (1915) "Kinematosk Opische Untersuchungen," *Zeitschrift fur Psychologie 72*, 193-296.

**Kuffler, S. W.** (1953) "Discharge Patterns and Functional Organization of Mammalian Retina," *Journal of Neurophysiology 16*, 37-68.

**Kulikowski, J. J. and King-Smith, P. E.** (1973) "Spatial Arrangement of Line, Edge and Grating Detectors Revealed by Subthreshold Summation," *Vision Research 13*, 1455-1478.

**Kulikowski, J. J. and Tolhurst, D. J.** (1973) "Psychophysical Evidence for Sustained and Transient Channels in Human Vision," *Journal of Physiology (London) 232*, 149-163.

**Lappin, J. S. and Bell, H. H.** (1976) "The Detection of Coherence in Moving Random Dot Patterns," *Vision Research 16*, 161-168.

**Lawton, D. T.** (1983) "Processing Translational Motion Sequences," *Computer Vision, Graphics, and Image Processing 22*, 116-144.

**Lee, D. N.** (1972) "Stimulus Pairing in Sequential Phi Motion," *Perception 1*, 85-91

**Leese, J. A., Novak, C. S. and Taylor, V. R.** (1970) "The Determination of Cloud Pattern Motion from Geosynchronous Satellite Image Data," *Pattern Recognition 2*, 279-292.

**Legge, G. E.** (1978) "Sustained and Transient Mechanisms in Human Vision: Temporal and Spatial Properties," *Vision Research 18*, 69-81.

**Levick, W. R., Cleland, B. G. and Dubin, M. W.** (1972) "Lateral Geniculate Neurons of Cat: Retinal Inputs and Physiology," *Investigative Opthalmology 11*, 302-311.

**Lillestrand, R. L.** (1972) "Techniques for Change Detection," *IEEE Transactions on Computers c-21*, 654-659.

**Limb, J. O. and Murphy, J. A.** (1975) "Estimating the Velocity of Moving Images in Television Signals," *Computer Graphics and Image Processing 4*, 311-327.

**Lin, C.-S., Weller, R. E. and Kaas, J. H.** (1982) "Cortical Connections of Striate Cortex in the Owl Monkey," *Journal of Comparative Neurology 211*, 165-176.

**Lit, A.** (1949) "The Magnitude of the Pulfrich Stereophenomenon as a Function of Binocular Differences of Intensity at Various Levels of Illumination," *American Journal of Psychology 62*, 159-181.

**Longuet-Higgins, H. C.** (1981) "A Computer Algorithm for Reconstructing a Scene From Two Projections," *Nature 293*, 133-135.

**Longuet-Higgins, H. C. and Prazdny, K.** (1981) "The Interpretation of Moving Retinal Images," *Proceedings of the Royal Society of London B. 208*, 385-397.

**Luenberger, D. G.** (1973) *Introduction to Linear and Nonlinear Programming*, Addison-Wesley, Reading, MA.

**Lund, J. S. and Boothe, R. G.** (1975) "Interlaminar Connections and Pyramidal Neuron Organization in the Visual Cortex, Area 17, of the Macaque Monkey," *Journal of Comparative Neurology 159*, 305-334.

**Lund, J. S., Henry, C. L., Macqueen, C. L. and Harvey, A. R.** (1979) "Anatomical Organization of the Primary Visual Cortex (Area 17) of the Cat. A Comparison with Area 17 of the Macaque Monkey," *Journal of Comparative Neurology 184*, 599-618.

**Lund, J. S., Lund, R. D., Hendrickson, A. E., Bunt, A. H. and Fuchs, A. F.** (1975) "The Origin of Efferent Pathways from the Primary Visual Cortex, Area 17, of the Macaque Monkey as Shown by Retrograde Transport of Horseradish Peroxidase," *Journal of Comparative Neurology 164*, 287-304.

**Marr, D.** (1976a) "Artificial Intelligence – A Personal View," *Artificial Intelligence 9* 37-48.

**Marr, D.** (1976b) "Early Processing of Visual Information," *Philosophical Transactions of the Royal Society of London B. 275*, 483-534.

**Marr, D.**, *VISION* (1982) W. H. Freeman Co., San Francisco.

**Marr, D. and Hildreth, E. C.** (1980) "Theory of Edge Detection," *Proceedings of the Royal Society of London B. 207*, 187-217.

**Marr, D. and Poggio, T.** (1977) "From Understanding Computation to Understanding Neural Circuitry," *Neuroscience Research Program Bulletin 15*, 470-488.

**Marr, D. and Poggio, T.** (1979) "A Computational Theory of Human Stereo Vision," *Proceedings of the Royal Society of London B. 204*, 301-328.

**Marr, D., Poggio, T. and Hildreth, E. C.** (1980) "Smallest Channel in Early Human Vision," *Journal of the Optical Society of America 70*, 868-870.

**Marr, D. and Ullman, S.** (1981) "Directional Selectivity and its Use in Early Visual Processing," *Proceedings of the Royal Society of London B. 211*, 151-180.

**Martinez-Millan, M. and Hollander, H.** (1975) "Cortico-Cortical Projections from Striate Cortex of the Squirrel Monkey *(Saimiri Sciureus)*. A Radiographic Study," *Brain Research 83*, 405-417.

**Mather, G. and Anstis, S. M.** (1983) "Theories of Motion Perception," presented at the ACM Interdisciplinary Workshop on Motion: Representation and Perception, Toronto, Canada.

**Maunsell, J. H. R. and Van Essen, D. C.** (1982) "The Connections of the Middle Temporal Visual Area in the Macaque Monkey," *Neuroscience Abstracts*, 811.

**Maunsell, J. H. R. and Van Essen, D. C.** (1983) "Functional Properties of Neurons in Middle Temporal Visual Area of the Macaque Monkey. I. Selectivity for Stimulus Direction, Speed and Orientation," *Journal of Neurophysiology 49*, 1127-1147.

**Mayhew, J. E. W. and Frisby, J. P.** (1981) "Psychophysical and Computational Studies Towards a Theory of Human Stereopsis," *Artificial Intelligence 17*, 349-385.

**McKee, S.** (1981) "A Local Mechanism for Differential Velocity Detection," *Vision Research 21*, 491-500.

**Miezin, F., McGuinness, E. and Allman, J.** (1982) "Antagonistic Direction Specific Mechanisms in Area MT in the Owl Monkey," *Neuroscience Abstracts*, 681.

**Montero, V. M.** (1980), "Patterns of Connections from the Striate Cortex to Cortical Visual Areas in Superior Temporal Sulcus of Macaque and Middle Temporal Gyrus of Owl Monkey," *Journal of Comparative Neurology 189*, 45-59.

**Morgan, M. J. and Ward, R.** (1980) "Conditions for Motion Flow in Dynamic Visual Noise," *Vision Research 20*, 431-435.

**Mountcastle, V. B., editor** (1974) *Medical Physiology*, C. V. Mosby Company, Saint Louis.

**Movshon, J. A.** (1975) "The Velocity Tuning of Single Units in Cat Striate Cortex," *Journal of Physiology (London) 249*, 445-468.

**Movshon, J. A., Adelson, E. H., Gizzi, M. S. and Newsome, W. T.** (1983) "The Analysis of Moving Visual Patterns," to appear in *Pattern Recognition Mechanisms*, ed. by C. Chagas, R. Gattas and C. G. Gross, Vatican Press, Rome. Also to appear in *Experimental Brain Research*.

**Mutch, K. M. and Thompson, W. B.** (1982) "Hierarchical Estimation of Spatial Properties from Motion," *University of Minnesota Computer Science Department Technical Report 82-22.*

**Nagel, H.-H.** (1981) "On the Derivation of 3D Rigid Point Configurations from Image Sequences," *Proceedings of the IEEE Conference on Pattern Recognition and Image Processing*, 103-108.

**Nagel, H.-H.** (1982) "On Change Detection and Displacement Vector Estimation in Image Sequences," *Pattern Recognition Letters 1*, 55-59.

**Nakayama, K.** (1981) "Differential Motion Hyperacuity Under Conditions of Common Image Motion," *Vision Research 21*, 1475-1482.

**Nakayama, K. and Tyler, C. W.** (1981) "Psychophysical Isolation of Movement Sensitivity by Removal of Familiar Position Cues," *Vision Research 21*, 427-433.

**Neff, W.** (1936) "A Critical Investigation of the Visual Apprehension of Movement," *Americal Journal of Psychology XLVIII*, 1-42.

**Neuhaus, W.** (1930) "Experimentelle Untersuchung der Scheinbewegung," *Archiv fur die Gesamte Psychologie 75*, 315-348.

**Newsome, W. T., Mikami, A. and Wurtz, R. H.** (1982) "Direction Selective Responses to Sequentially Flashed Stimuli in Extrastriate Area MT in the Awake Macaque Monkey," *Neuroscience Abstracts*, 812.

**Orban, G. A., Kennedy, H. and Maes, H.** (1981) "Response to Movement of Neurons in Areas 17 and 18 of the Cat: Velocity Sensitivity," *Journal of Neurophysiology 45*, 1043-1058.

**Pantle, A. J. and Picciano, L.** (1976) "A Multistable Display: Evidence for Two Separate Motion Systems in Human Vision," *Science 193*, 500-502.

**Peichl, L. and Wassle, H.** (1979) "Size, Scatter and Convergence of Ganglion Cell Receptive Field Centers in the Cat Retina," *Journal of Physiology (London) 291*, 117-141.

**Petersen, S. E., Robinson, D. L. and Keys, W.** (1982) "A Physiological Comparison of the Lateral Pulvinar and Area 7 in the Behaving Macaque," *Neuroscience Abtracts*, 681.

**Petersik, J. T.** (1980) "The Effect of Spatial and Temporal Factors on the Perception of Stroboscopic Rotation Stimulations," *Perception 9*, 271-283.

**Petersik, J. T., Hicks, K. I. and Pantle, A. J.** (1978) "Apparent Movement of Successively Generated Subjective Patterns," *Perception 7*, 371-383.

**Petersik, J. T. and Pantle, A.** (1979) "Factors Controlling the Competing Sensations Produced by a Bistable Stroboscopic Motion Display," *Vision Research 19*, 143-154.

**Poggio, G. F., Doty, R. W. and Talbot, W. H.** (1977) "Foveal Striate Cortex of Behaving Monkey: Single-neuron Response to Square-wave Gratings During Fixation of Gaze," *Journal of Neurophysiology 40*, 1369-1391.

**Poggio, T.** (1983) "Visual Algorithms," in: *Physical and Biological Processing of Images*, ed. by O. J. Braddick and A. C. Sleigh, Springer-Verlag, Berlin.

**Poggio, T. and Reichardt, W.** (1973) "Considerations on Models of Movement Detection," *Kybernetik 13*, 223-277.

**Poggio, T. and Reichardt, W.** (1976) "Visual Control of Orientation Behavior in the Fly," *Quarterly Reviews of Biophysics 9*, 377-438.

**Poggio, T. and Reichardt, W.** (1980) "On the Representation of Multi-Input Systems: Computational Properties of Polynomial Algorithms," *Biological Cybernetics 37*, 167-186.

**Poggio, T., Reichardt, W. and Hausen, K.** (1981) "A Neuronal Circuitry for Relative Movement Discrimination by the Visual System of the Fly," *Naturwissenschaften 68*, 443-446.

**Potter, J. L.** (1975) "Velocity as a Cue to Segmentation," *IEEE Transactions on Systems, Man, and Cybernetics SMC-5*, 390-394.

**Potter, J. L.** (1977) "Scene Segmentation Using Motion Information," *Computer Graphics and Image Processing 6*, 558-581.

**Pratt, W.** (1978) *Digital Image Processing*, Wiley and Sons, New York.

**Prazdny, K.** (1980) "Egomotion and Relative Depth Map from Optical Flow," *Biological Cybernetics 36*, 87-102.

**Prazdny, K.** (1983) "On the Information in Optical Flows," *Computer Vision, Graphics, and Image Processing 22*, 239-259.

**Ramachandran, V. S. and Anstis, S. M.** (1983) "Extrapolation of Motion Path in Human Visual Perception," *Vision Research 23*, 83-85.

**Regan, D. and Spekreijse, H.** (1970) "Electrophysiological Correlate of Binocular Depth Perception in Man," *Nature 225*, 92-94.

**Reichardt, W.** (1961) "Autocorrelation, a Principle for the Evaluation of Sensory Information by the Central Nervous System," in: *Sensory Communication*, ed. by W. A. Rosenblith, MIT Press, Cambridge.

**Reichardt, W. and Poggio, T.** (1976) "Visual Control of Orientation Behavior in the Fly, Part I. A Quantitative Analysis," *Quarterly Review of Biophysics 9*, 311-375.

**Reichardt, W. and Poggio, T.** (1979) "Figure-ground Discrimination by Relative Movement in the Visual System of the Fly," *Biological Cybernetics 35*, 81-100.

**Reichardt, W., Poggio, T. and Hausen, K.** (1983) "Figure-Ground Discrimination by Relative Movement in the Visual System of the Fly. Part II: Towards the Neural Circuitry," *Biological Cybernetics 46*, 1-30.

**Richards, W. and Lieberman, H. R.** (1982) "Velocity Blindness During Shearing Motion," *Vision Research 22*, 97-100.

**Richter, J.** (1982) *Association for Research in Vision and Ophtalmology, Abstracts.*

**Richter, J. and Ullman, S.** (1980) "A Model for the Spatio-temporal Organization of X and Y-type Ganglion Cells in the Primate Retina," *MIT Artificial Intelligence Laboratory Memo 573.*

**Richter, J. and Ullman, S.** (1982) "A Model for the Temporal Organization of X-type and Y-type Receptive Fields in the Primate Retina," *Biological Cybernetics 43*, 127-145.

**Roach, J. W. and Aggarwal, J. K.** (1979) "Computer Tracking of Objects Moving in Space," *IEEE Transactions on Pattern Analysis and Machine Intelligence PAMI-1*, 127-135.

**Rodieck, R. W.** (1965) "Quantitative Analysis of Cat Retinal Ganglion Cell Responses to Visual Stimuli," *Vision Research 5*, 583-601.

**Rodieck, R. W.** (1967) "Receptive Fields in the Cat Retina: A New Type," *Science 157*, 90-92.

**Rodieck, R. W.** (1979) "Visual Pathways," *Annual Reviews of Neuroscience 2*, 193-225.

**Rodieck, R. W. and Stone, J.** (1965) "Analysis of Receptive Fields of Cat Retinal Ganglion Cells," *Journal of Neurophysiology 28*, 833-849.

**Rosenfeld, A. and Kak, A.** (1976) *Digital Picture Processing*, Academic Press, New York.

**Rudin, W.** (1966) *Real and Complex Analysis*, McGraw-Hill Co., New York.

**Rudin, W.** (1973) *Functional Analysis*, McGraw-Hill Co., New York.

**Sachs, M. B., Nachmias, J. and Robson, J. G.** (1971) "Spatial Frequency Channels in Human Vision," *Journal of the Optical Society of America 61*, 1176-1186.

**Schiller, P. H.** (1982) "Central Connections of the Retinal ON and OFF Pathways," *Nature 297*, 580-583.

**Schiller, P. H., Finlay, B. L. and Volman, S. F.** (1976a) "Quantitative Studies of Single-cell Properties in Monkey Striate Cortex. I. Spatiotemporal Organization of Receptive Fields," *Journal of Neurophysiology 39*, 1288-1319.

**Schiller, P. H., Finlay, B. L. and Volman, S. F.** (1976b) "Quantitative Studies of Single-cell Properties in Monkey Striate Cortex. II. Orientation Specificity and Ocular Dominance," *Journal of Neurophysiology 39*, 1320-1333.

**Schiller, P. H. and Malpeli, J. G.** (1977) "Properties and Tectal Projections of Monkey Retinal Ganglion Cells," *Journal of Neurophysiology 40*, 428-445.

**Schiller, P. H. and Malpeli, J. G.** (1978) "Functional Specificity of Lateral Geniculate Nucleus Lamina of the Rhesus Monkey," *Journal of Neurophysiology 41*, 788-797.

**Schulman, P. H.** (1979) "A New Motion Illusion," *Perception 8*, 659-664.

**Schunck, B. G.** (1983) "Motion Segmentation and Estimation," *PhD Thesis, MIT Department of Electrical Engineering and Computer Science.*

**Schunck, B. G. and Horn, B. K. P.** (1981) "Constraints on Optical Flow Computation," *Proceedings of the IEEE Conference on Pattern Recognition and Image Processing*, August, 205-210.

**Shepard, R. N. and Judd, S. A.** (1976) "Perceptual Illusion of Rotation of Three-dimensional Objects," *Science 191*, 952-954.

**Sherman, S. M., Wilson, J. R., Kaas, J. H. and Webb, S. V.** (1976) "X- and Y-cells in the Dorsal Lateral Geniculate Nucleus of the Owl Monkey (Aotus Trivergatus)," *Science 192*, 475-477.

**Singer, W. and Creutzfeldt, O. D.** (1970) "Reciprocal Lateral Inhibition of ON- and OFF-center Neurons in the Lateral Geniculate Body of the Cat," *Experimental Brain Research 10*, 311-330.

**Smith, E. A. and Phillips, D. R.** (1972) "Automated Cloud Tracking Using Precisely Aligned Digital ATS Pictures," *IEEE Transactions on Computers c-21*, 715-729.

**Spatz, W. B.** (1977) "Topographicallly Organized Reciprocal Connections Between Areas 17 and MT (Visual Area of Superior Temporal Sulcus) in Marmoset, *Callithrixjacchus*," *Experimental Brain Research 27*, 91-108.

**Spitzberg, R. and Richards, W. A.** (1975) "Broad Band Spatial Filters in the Human Visual System," *Vision Research 15*, 837-841.

**Standage, G. P. and Benevento, L. A.** (1982) "Lack of Dorsal Lateral Geniculate Input to Extrastriate Cortical Areas MT and Dorsal Visual 2 in the Macaque Monkey," *Neuroscience Abstracts*, 680.

**Stone, J. and Fabian, M.** (1966) "Specialized Receptive Fields of the Cat's Retina," *Science 152*, 1277-1279.

**Stone, J. and Hoffman, K-P.** (1972) "Very Slow Conducting Ganglion Cells in the cat's Retina: A Major. New Functional Type," *Brain Research 43*, 610-616.

**Ternus, J.** (1926) "Experimentelle Untersuchung uber Phanomenale Identitat," *Psychologische Forschung 7* 81-136. Translated in: *A Source Book of Gestalt Psychology*, ed. by W. D. Ellis, Humanities Press, New York, 1967.

**Terzopoulos, D.** (1982) "Multi-level Reconstruction of Visual Surfaces: Variational Principles and Finite Element Representations," *MIT Artificial Intelligence Laboratory Memo 671*.

**Thompson, W. B.** (1980) "Combining Motion and Contrast for Segmentation," *IEEE Transactions on Pattern Analysis and Machine Intelligence PAMI-2*, 543-549.

**Thompson, W. B. and Barnard, S. T.** (1981) "Lower-level Estimation and Interpretation of Visual Motion," *IEEE Computer*, August, 20-28.

**Thompson, W. B., Mutch, K. M. and Berzins, V.** (1982) "Edge Detection in Optical Flow Fields," *Proceedings of the Second National Conference on Artificial Intelligence*, August, 26-29.

**Tigges, J., Spatz, W. B. and Tigges, M.** (1974) "Efferent Cortico-Cortical Fiber Connections of Area 18 in the Squirrel Monkey *(Samiri)*," Journal of Comparative Neurology 158, 219-236.

**Tigges, J., Tigges, M., Anschel, S., Cross, N. A., Letbetter, W. D. and McBride, R. L.** (1981) "Areal and Laminar Distribution of Neurons Interconnecting the Central Visual Cortical Areas 17, 18, 19, and MT in Squrrel Monkey (Saimiri)," *Journal of Comparative Neurology 202*, 539-560.

**Tolhurst, D. J.** (1973) "Separate Channels for the Analysis of the Shape and the Movement of a Moving Visual Stimulus," *Journal of Physiology (London) 231*, 385-402.

**Tolhurst, D. J.** (1975) "Sustained and Transient Channels in Human Vision," *Vision Research 15*, 1151-1155.

**Tolhurst, D. J. and Movshon, J. A.** (1975) "Spatial and Temporal Contrast Sensitivity of Striate Cortical Neurons," *Nature 257*, 674-675.

**Tolhurst, D. J., Movshon, J. A. and Thompson, I. D.** (1981) "The Dependence of Response Amplitude and Variance of Cat Visual Cortical Neurones on Stimulus Contrast," *Experimental Brain Research 41*, 414-419.

**Torre, V. and Poggio, T.** (1978) "A Synaptic Mechanism Possibly Underlying Directional Selectivity to Motion," *Proceedings of the Royal Society of London B. 202*, 409-416.

**Tsai, R. Y. and Huang, T. S.** (1981) "Uniqueness and Estimation of Three-dimensional Motion Parameters of Rigid Objects With Curved Surfaces," *University of Illinois at Urbana-Champaign, Coordinated Science Laboratory Report R-921.*

**Ullman, S.** (1979a) *The Interpretation of Visual Motion*, MIT Press, Cambridge and London.

**Ullman, S.** (1979b) "Relaxation and Constrained Optimization by Local Processes," *Computer Graphics and Image Processing 9*, 115-125.

**Ullman, S.** (1980) "The Interpretation of Three-dimensional Structure from Motion," *Proceedings of the Royal Society of London B. 203*, 405-426.

**Ullman, S.** (1981) "Analysis of Visual Motion by Biological and Computer Systems," *IEEE Computer*, August, 57-69.

**Ullman, S.** (1983a) "Computational Studies in the Interpretation of Structure and Motion: Summary and Extension," *MIT Artificial Intelligence Laboratory Memo 706.*

**Ullman, S.** (1983b) "Maximizing Rigidity: The Incremental Recovery of 3-D Structure From Rigid and Rubbery Motion," *MIT Artificial Intelligence Laboratory Memo 721.*

**Ullman, S. and Hildreth, E. C.** (1983) "The Measurement of Visual Motion," in: *Physical and Biological Processing of Images*, ed. by O. J. Braddick and A. C. Sleigh, Springer-Verlag, Berlin.

**Ungerleider, L. G., Desimone, R. and Mishkin, M.** (1982) "Cortical Projections of Area MT in the Macaque," *Neuroscience Abstracts*, 680.

**Ungerleider, L. G. and Mishkin, M.** (1979) "The Striate Projection Zone in the Superior Temporal Sulcus of Macaca Mulatta: Location and Topographic Organization," *Journal of Comparative Neurology 188*, 347-366.

**Van Essen, D. C., Maunsell, J. H. R. and Bixby, J. L.** (1981a) "The Middle Temporal Visual Area in the Macaque: Myeloarchitecture, Connections, Functional

Properties and Topographical Organization," *Journal of Comparative Neurology 199*, 293-326.

**Van Essen, D. C., Maunsell, J. H. R. and Bixby, J. L.** (1981b) "The Organization of Extrastriate Visual Areas in the Macaque Monkey," in: *Multiple Cortical Areas*, ed. by C. N. Woolsey, Humana Press, Englewood Cliffs, NJ.

**Wallach, H.** (1976) "On Perceived Identity: 1. The Direction of Motion of Straight Lines", in: *On Perception*, ed. by H. Wallach, Quadrangle, New York.

**Wallach, H. and O'Connell, D. N.** (1953) "The Kinetic Depth Effect," *Journal of Experimental Psychology 45*, 205-217.

**Wallach, H., O'Leary, A. and McMahon, M. L.** (1982) "Three Stimuli for Visual Motion Perception Compared," *Perception and Psychophysics 32*, 1-6.

**Wallach, H., Weisz, A. and Adams, P. A.** (1956) "Circles and Derived Figures in Rotation," *American Journal of Psychology 69*, 48-59.

**Wassle, H., Levick, W. R. and Cleland, B. G.** (1975) "The Distribution of the Alpha Type of Ganglion Cells in the Cat's Retina," *Journal of Comparative Neurology 159*, 419-438.

**Watson, A. B.** (1982) "Summation of Grating Patches Indicates Many Types of Detector at One Retinal Location," *Vision Research 22*, 17-25.

**Watson, A. B.** (1983) "Detection and Recognition of Simple Spatial Forms," in: *Physical and Biological Processing of Images*, ed. by O. J. Braddick and A. C. Sleigh, Springer-Verlag, Berlin.

**Watson, A. B. and Nachmias, J.** (1977) "Patterns of Temporal Interaction in the Detection of Gratings," *Vision Research 17*, 893-902.

**Watt, R. J. and Morgan, M. J.** (1982) "Mechanisms Responsible for the Assessment of Visual Location: Theory and Evidence," *Vision Research 23*, 97-109.

**Watt, R. J. and Morgan, M. J.** (1983) "The Recognition and Representation of Edge Blur: Evidence for Spatial Primitives in Human Vision," *in preparation.*

**Weller, R. E., Graham, J. and Kaas, J. H.** (1979) "Cortical Connections of Striate Cortex in Macaque Monkeys," *Association for Research in Vision and Ophthalmology, Abtracts*, 157.

**Weller, R. E. and Kaas, J. H.** (1978) "Connections of Striate Cortex With the Posterior Bank of the Superior Temporal Sulcus in Macaque Monkeys," *Neuroscience Abtracts 4*, 650.

**Weller, R. E. and Kaas, J. H.** (1981) "Cortical and Subcortical Connections of Visual Cortex in Primates," in: *Multiple Cortical Areas*, ed. by C. N. Woolsey, Humana Press, Englewood Cliffs, NJ.

**Wertheimer, M.** (1912) "Experimentelle Studien Uber Das Sehen Von Bewegung," *Zeitchrift fur Psychologie 61*, 161-265.

**Wilson, H. R.** (1980) "Spatiotemporal Characterization of a Transient Mechanism in the Human Visual System," *Vision Research 20*, 443-452.

**Wilson, H. R.** (1983) "Psychophysical Evidence for Spatial Channels," in: *Physical and Biological Processing of Images*, ed. by O. J. Braddick and A. C. Sleigh, Springer-Verlag, Berlin.

**Wilson, H. R. and Bergen, J. R.** (1979) "A Four Mechanism Model for Threshold Spatial Vision," *Vision Research 19*, 19-32.

**Wilson, H. R. and Giese, S. C.** (1977) "Threshold Visibility of Frequency Gradient Patterns," *Vision Research 17*, 1177-1190.

**Wilson, H. R.** (1980) "Spatiotemporal Characterization of a Transient Mechanism in the Human Visual System," *Vision Research 5*, 443-552.

**Wohlgemuth, A.** (1911) *British Journal of Psychology*, monograph supplement 1.

**Wolferts, K.** (1974) "Special Problems in Interactive Image Processing for Traffic Analysis," *Proceedings of the Second International Joint Conference on Pattern Recognition*, 1-2.

**Wright, M.** (1982) *Practical Optimization*, Academic Press: New York.

**Yuille, A. L.** (1983) "The Smoothest Velocity Field and Token Matching Schemes," *MIT Artificial Intelligence Laboratory Memo 724.*

**Zeeman, W. P. C. and Roelofs, C. O.** (1953) "Some Aspects of Apparent Motion," *Acta Psychologia 9*, 159-181.

**Zeki, S. M.** (1974) "Functional Organization of a Visual Area in the Posterior Bank of the Superior Temporal Sulcus of the Rhesus Monkey," *Journal of Physiology (London) 236*, 549-573.

**Zeki, S. M.** (1978) "Uniformity and Diversity of Structure and Function in Rhesus Monkey Prestriate Visual Cortex," *Journal of Physiology (London) 277*, 273-290.

**Zeki, S. M.** (1980) "The Response Properties of Cells in the Middle Temporal Area (Area MT) of Owl Monkey Visual Cortex," *Proceedings of the Royal Society of London B. 207*, 239-248.

# Author Index

# Subject Index